Engineering Applications of Computational Methods

Volume 22

The book series Engineering Applications of Computational Methods addresses the numerous applications of mathematical theory and latest computational or numerical methods in various fields of engineering. It emphasizes the practical application of these methods, with possible aspects in programming. New and developing computational methods using big data, machine learning and AI are discussed in this book series, and could be applied to engineering fields, such as manufacturing, industrial engineering, control engineering, civil engineering, energy engineering and material engineering.

The book series Engineering Applications of Computational Methods aims to introduce important computational methods adopted in different engineering projects to researchers and engineers. The individual book volumes in the series are thematic. The goal of each volume is to give readers a comprehensive overview of how the computational methods in a certain engineering area can be used. As a collection, the series provides valuable resources to a wide audience in academia, the engineering research community, industry and anyone else who are looking to expand their knowledge of computational methods.

This book series is indexed in both the **Scopus** and **Compendex** databases.

Hongpeng Yin · Han Zhou · Yi Chai · Qiu Tang

Data-Driven Fault Diagnosis for Complex Industrial Processes

Towards Fault Prediction, Detection and Identification

Springer

Hongpeng Yin
College of Automation
Chongqing University
Chongqing, China

Yi Chai
College of Automation
Chongqing University
Chongqing, China

Han Zhou
Peng Cheng Laboratory
Shenzhen, China

Qiu Tang
Shandong University Scientific Journals Press
Shandong University
Jinan, Shandong, China

ISSN 2662-3366 ISSN 2662-3374 (electronic)
Engineering Applications of Computational Methods
ISBN 978-981-96-3152-0 ISBN 978-981-96-3153-7 (eBook)
https://doi.org/10.1007/978-981-96-3153-7

This work was supported by National Natural Science Foundation of China (U2034209, 62273062, 61773080) and Natural Science Foundation of Chongqing Municipality (2022NSCQ-LZX0324, cstc2021ycjh-bgzxm0028).

This Springer imprint is published by the registered company Springer Nature Singapore Pte Ltd.
The registered company address is: 152 Beach Road, #21-01/04 Gateway East, Singapore 189721, Singapore

Preface

In modern industrial processes, where efficiency, productivity, and safety are of utmost importance, the pursuit of fault diagnosis has become more crucial than ever. The widespread use of computer systems and advancements in sensor technology generate vast amounts of real-time and historical process data. This has raised questions about how to leverage these data sources to enhance efficiency and create a more resilient and sustainable future for industrial process fault diagnosis. Consequently, this work focuses on data-driven fault diagnosis methods tailored to the unique demands of industrial processes.

While existing books have long contributed to fault diagnosis, the application of data-driven methods to dynamic, complex industrial processes is still an emerging area of research. This book, shaped by our collaboration with the automotive and process industries, focuses on the development of data-driven fault diagnostic methods specifically for industrial processes. Data-driven fault diagnosis stands at the intersection of engineering and data science. Drawing from data science, machine learning, and domain-specific expertise, this book summarizes our work on the past decades of research in fault prediction, detection, and identification. It includes foundational theories in the diagnosis of industrial process faults alongside recent advances in the field. Each theory and methodology is supported by experimental validations or applications across various industrial domains.

The outline of this book is listed below. This book starts with an introduction, definition, and related work of fault diagnosis in industrial processes (Chap. 1). Followed by the first technical part (Chap. 2), it presents methods for predicting health states and forecasting faults in dynamic, time-varying industrial processes. The second technical part (Chaps. 3–5) centers on fault detection, exploring various discriminant monitoring methods for nonlinear industrial processes. The third technical part (Chaps. 6–9) addresses methods for identifying faults in industrial processes characterized by multiple operational conditions, and non-stationarity. Finally, Chap. 10 gives a unified framework for fault diagnosis with detection and identification ability.

Our goal is to provide a valuable resource for industrial practitioners, researchers, and academicians. We believe that the strategies and knowledge shared within these pages will not only contribute to the field of fault diagnosis but also inspire future innovations in dynamic industrial environments.

Chongqing, China
October 2024

Hongpeng Yin

Acknowledgements This book is the result of years of our research and collaboration, and I am deeply grateful to the many individuals and institutions whose support has made this work possible.

The contents of this book were mostly developed at Chongqing University, P. R. China. First and foremost, I would like to express my deepest appreciation to Chongqing University for providing an enriching academic environment where this research could thrive. I am particularly grateful to the School of Automation for its continued support. The resources and facilities available here played a crucial role in the development and completion of this book.

I extend my deepest appreciation to the various foundations whose funding made this research possible. Their generous support played a crucial role in expanding the scope of data-driven fault diagnosis, enabling the development of the extensive research presented in this book. Over the past years, the following grants have contributed significantly to research materials and findings: the National Key Research and Development Program of China (2023YFB3308800), the National Natural Science Foundation of China (U2034209, 62273062, 61773080, 61633005, 61374135, 60974090, 62303276), the Natural Science Foundation of Chongqing (CSTB2022NSCQ-LZX0324, CSTC2015JCYJB0569), Chongqing Talents: Exceptional Young Talents Project (CSTC2021YCJH-BGZXM0028), Shandong Provincial Natural Science Foundation (ZR2022QF031), the Fundamental Funds for the Central Universities (2023CDJKYJH029, 2020CDCGZDH066, 2019CDCGZDH336, cqu2018CDHB1B04).

I express sincere gratitude to our colleagues Dr. Yan Qin, Dr. Jianfeng Qu, Dr. Shanbi Wei, Dr. Jingdong Lin, Dr. Yanxia Li, whose constructive suggestions significantly improved the manuscript. We also thank our students at Chongqing University, Dandan Zhao, Li Cai, Renpeng Mo, Yudi Zhang, Peng Zhang, Jiajun Chen, RuiWang, Tao Zhou, YuxuanWu, Qiuhao Lu, Xinyi Yang, Zicheng Huang for their contributions of numerous research ideas over the years. I would also like to acknowledge the editorial team at Springer for their professionalism and dedication in bringing this manuscript to publication. Their attention and feedback improved the quality of this book significantly.

Lastly, I extend my heartfelt gratitude to my family and friends for their constant encouragement and patience throughout this journey. Their support provided me with the motivation to persevere during the most challenging times.

To everyone who contributed to this work, I offer my sincere thanks. Your contributions, directly or indirectly, are deeply valued and greatly appreciated.

Chongqing, China
October 2024

Hongpeng Yin

Competing Interests The authors have no competing interests to declare that are relevant to the content of this manuscript.

Contents

Abbreviations

AEDL	Auto-encoder Embedded Dictionary Learning
AIC	Akaike Information Criterion
ANN	Artificial Neural Networks
CCR	Cumulative Contribution Rate
C-MAPSS	Commercial Modular Aero-Propulsion System Simulation
CMFP	Cranfield Multi-phase Flow Process
CNN	Convolutional Neural Networks
CSTR	Continuous Stirred Tank Reactor
CVA	Canonical Variate Analysis
DBN	Deep Belief Network
FAR	False Alarm Rate
FBFP	Fed-Batch Fermentation Penicillin
FCM	Fuzzy C-Means
FD	Fault diagnosis
FDA	Fisher Discriminant Analysis
FDGLPP	Fisher Discriminant Global-Local Preserving Projection
FDR	Fault Detection Rate
FSFDP	Fast Search and Find of Density Peaks
GAN	Generative Adversarial Network
GMM	Gaussian Mixture Model
GPR	Gaussian Process Regression
GRU	Gated Recurrent Unit
HI	Health Indices
HMM	Hidden Markov Model
IAP	Indirect Attribute Prediction
ICA	Independent Component Analysis
IoT	Internet of Things
ISVDD	Incremental Support Vector Data Description
KDE	Kernel Density Estimation
KKT	Karush-Kuhn-Tucker
KLD	Kullback-Leibler Divergence

KM	K-Means
KNN	K-Nearest Neighbors
KPCA	Kernel Principal Component Analysis
LDA	Linear Discriminant Analysis
LPP	Local Preserving Projection
LSTM	Long Short Term Memory
MAE	Mean Absolute Error
MAPE	Mean Absolute Percentage Error
MAR	Miss Alarm Rate
MPF	Multi-Phase Flow
MSPM	Multivariate Statistical Process Monitoring
MSVDD	Multi-class Support Vector Data Description
NASA	National Aeronautics and Space Administration
NMF	Non-negative Matrix Factorization
NNSE	Neural Network Squared Exponential
PA	Passive Aggressive
PCA	Principal Component Analysis
PHM	Prognostics and Health Management
PID	Proportional-Integral-Derivative
PLS	Partial Least Squares
PPV	Positive Predictive Value
RF	Random Forests
RKMM	Robust K-Multiple-Mean
RMSE	Root Mean Square Error
RUL	Remaining Useful Life
SAE	Stacked Auto-Encoders
SEDA	Sparse Exponential Discriminant Analysis
SVD	Singular Value Decomposition
SVDD	Support Vector Data Description
TEP	Tennessee Eastman Process
TPR	True Positive Rate
ZSFD	Zero-Sample Fault Diagnosis

Chapter 1
Introduction

1.1 Background and Definitions

Industrial processes span various fields, including chemicals, petroleum and natural gas, and power generation. It usually involves a series of operations that transform, process, or combine raw materials, energy, and other resources to produce a final product. These processes. Typically, an industrial process comprises a sequence of interconnected operational units or stages, each designed with specific tasks and functions to handle particular flows of material, energy, or information. These units operate in a designated order, forming a continuous workflow.

In recent years, with evolving market demands and the drive for enhanced production efficiency and competitiveness, new industrial production techniques and methods have emerged, making industrial processes larger in scale, more complex, and increasingly intelligent. However, these advancements present significant challenges for maintaining and managing the health of these systems. Faults in modern industrial processes, if not promptly detected and addressed, can propagate and amplify, leading to systemic chain reactions and, potentially, major safety incidents. Such incidents pose severe risks, including personal injury, economic and property losses, and environmental pollution.

A fault in an industrial process is defined as any deviation of one or more system characteristics or parameters from their expected trajectory [1]. Fault-induced catastrophic incidents in industrial settings are not uncommon. For instance, in July 1988, the Piper Alpha oil platform[1] in the North Sea experienced a massive explosion during the startup of a liquefied natural gas pump. This disaster resulted in the deaths of 167 employees, the destruction of 300 platforms, and financial losses amounting to $3.4 billion. Subsequent investigations revealed that the explosion was caused by the failure of the pump's main safety valve and improper maintenance of the backup valve, which led to a condensate leak. In May 2020, a chemical plant in Visakhapat-

[1] Website: https://www.thechemicalengineer.com/features/piper-alpha-the-disaster-in-detail/ [Accessed in Oct. 2024].

H. Yin et al., *Data-Driven Fault Diagnosis for Complex Industrial Processes*, Engineering Applications of Computational Methods 22,
https://doi.org/10.1007/978-981-96-3153-7_1

nam,[2] India, suffered a styrene gas leak due to a fault in the cooling unit, resulting in 11 fatalities and over 2,000 poisonings. More recently in 2022, an explosion at the Shanghai Jinshan Petrochemical plant[3] claimed one life and injured several others, incurring losses estimated at tens of millions of yuan. Investigations attributed the incident to the aging and fracture of a clamp on the ethylene oxide purification tower, resulting in a leak and subsequent ignition of ethylene oxide.

The purpose of fault diagnosis is to accurately predict, detect and identify the unexpected deviation within a system [2]. Timely and precise fault diagnosis is essential for ensuring the reliability and safety of industrial processes. Diagnosis results provide operational staff with critical information for maintenance and repair decisions, helping to prevent potential issues, minimize production interruptions, and reduce equipment damage. This enables continuous, stable industrial operations, enhances product quality and production efficiency, lowers production costs, and strengthens overall competitiveness. Consequently, research into fault diagnosis technology to improve safety and sustainable development in industrial processes has become a significant focus in recent years, bearing substantial practical importance.

Fault diagnosis can be divided into three main tasks, as shown in Fig. 1.1:

- **Fault Prediction**: Forecast potential faults before they occur, allowing for proactive measures to prevent failures.
- **Fault Detection**: Identifying the presence of a fault in real-time, providing immediate alerts when a system deviates from normal operation. This task is also called fault monitoring.
- **Fault Identification**: Pinpointing the specific type, location, and cause of a detected fault to facilitate accurate troubleshooting and repair.

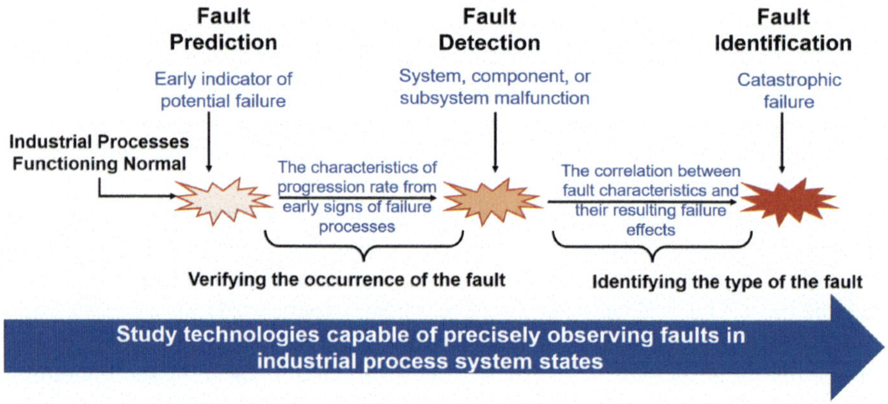

Fig. 1.1 The flowchart of fault diagnosis

[2] https://timesofindia.indiatimes.com/city/visakhapatnam/glitch-in-refrigeration-unit-led-to-vizag-gas-leak-official/articleshow/75604393.cms [Accessed in Oct. 2024].

[3] https://www.cnr.cn/shanghai/tt/20230421/t20230421_526226816.shtml *(In Chinese)*, [Accessed in Oct. 2024]).

Traditional fault diagnosis methods primarily rely on manual "regular inspections and reactive repairs." This approach has significant drawbacks: It requires extensive human resources for periodic checks and troubleshooting, resulting in high labor costs. Additionally, human factors can introduce subjectivity and inconsistency in inspections, increasing the risk of missed or false detections. In many industrial settings, the working environment may be particularly challenging, with conditions such as high altitudes, confined spaces, loud noise, and strong vibrations, which can impact worker health and safety. With advances in communication, computing, big data, and the Internet of Things (IoT) driven by the Fourth Industrial Revolution, modern fault diagnosis is increasingly shifting towards digitization, informatization, and intelligent automation.

1.2 Literature Review

As shown in Fig. 1.2, fault diagnosis techniques are generally classified into three categories: knowledge-based, model-based, and data-driven approaches. Knowledge-based methods, developed since the 1960s, rely on expert knowledge, rule-based systems, and heuristic techniques to identify and diagnose faults. Model-based methods, which emerged subsequently, use mathematical and physical models of systems to detect faults by identifying deviations from expected behavior. With advancements in computer systems and sensor technology, large volumes of real-time and historical process data are now available, enabling the rise of data-driven fault diagnosis. This approach leverages data mining techniques, including statistical analysis and deep learning, to extract patterns and diagnose faults.

In the following sections, this chapter will provide a thorough exploration of various fault diagnosis methods applicable to complex industrial processes, emphasizing fault prediction, detection, and identification. Each category will be analyzed in detail to clarify its core principles, contextual background, and the current state of research in the field of fault diagnosis.

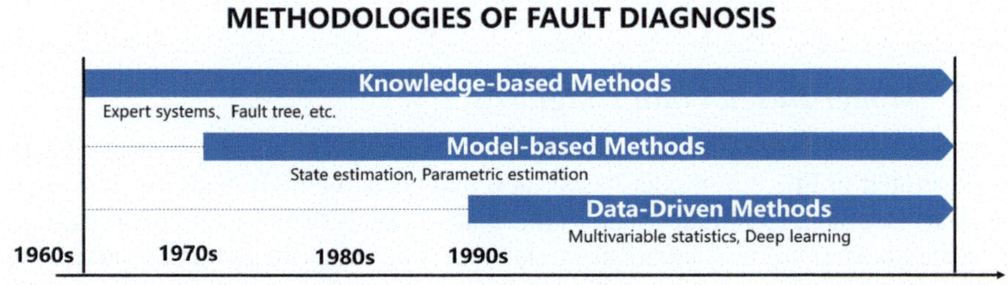

Fig. 1.2 Road map of fault diagnosis techniques for industrial processes

1.2.1 Knowledge-Based Fault Diagnosis

Knowledge-based fault diagnosis methods formalize expert knowledge and prior information to qualitatively characterize different aspects of faults in industrial systems, including fault locations, types, and propagation patterns. Deductive strategies can then be applied for fault diagnosis. Common approaches in this category include expert systems [3] and fault tree analysis [4].

Expert systems typically use knowledge representation to compile extensive expert-level knowledge and experience in a specific domain. They employ knowledge reasoning techniques to simulate the human fault diagnosis process. For example, Qian et al. [5] developed an intelligent expert system for fluid catalytic cracking units in chemical processes. This system utilizes IF-AND-THEN statements to store experiential knowledge about normal operating temperatures of various components. It then simulates the decision-making process of human experts using IF-ELSE statements to identify potential fault indicators during operation.

Fault trees are inverted tree diagrams that represent causal relationships using Boolean logic. They describe the causal links from undesirable base states to the top-level failure event, thereby determining fault occurrence rates. Lavasani et al. [6] created a fault tree with eight levels, including basic events, failure probabilities, and top events, specifically for petrochemical processes. This approach utilized fuzzy reasoning to quantitatively capture expert knowledge, clarifying fault propagation principles and enhancing diagnostic accuracy.

Other research on knowledge-based fault diagnosis methods has also been conducted by various teams, including Venkatasubramanian [7] for chemical processes, Bai [8] for rotating machinery.

Knowledge-based fault diagnosis methods are highly interpretable and are well-suited for smaller systems with relatively simple structures and known fault mechanisms. However, the effectiveness of these methods relies on the accumulation of knowledge and the expertise of the individuals involved. In the context of large, complex industrial processes characterized by multi-layered structures, strong coupling, and uncertainty, these fault diagnosis methods often struggle to provide comprehensive solutions.

1.2.2 Model-Based Fault Diagnosis

As illustrated in Fig. 1.3, model-based fault diagnosis methods begin by analyzing the system through techniques such as mechanistic analysis or system identification to create an accurate mathematical model. By comparing the actual output of the system with that of the mathematical model, these methods can identify discrepancies between expected and actual performance. Subsequently, inference techniques are used for pattern recognition to determine details such as fault type, location, and magnitude. Common approaches in this category include state estimation, parameter

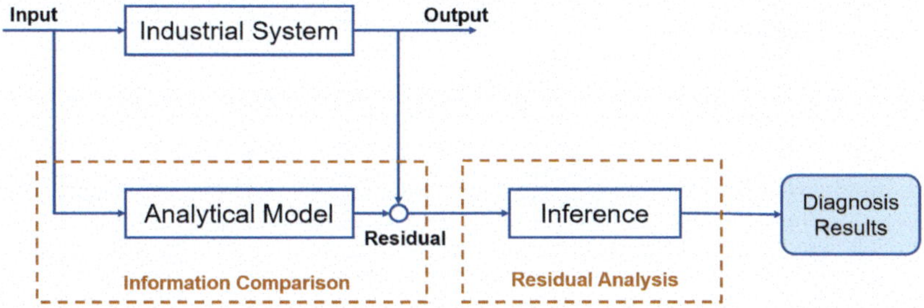

Fig. 1.3 The illustration of model-based fault diagnosis

estimation, and parity space methods. Systematic introductions and reviews of these methods can be found in references [1, 9].

State estimation involves comparing estimated values of observable parameters with actual measurements using designed observers when system parameters are fully known, allowing for fault identification. When system parameters are partially known or unknown, parameter estimation methods based on system identification techniques can be employed. This approach assumes that faults are reflected in changes to system parameters, allowing for fault identification through the estimation of corresponding physical parameter variations from input-output signals. Common parameter estimation methods include least squares [10] and regression analysis [11]. The parity space method involves designing an equivalent model for the system and determining system residuals by comparing the outputs of the equivalent model with actual outputs. Classification and discrimination techniques are then used to identify system faults. An early review for parity space-based fault diagnosis can be found in [12].

Model-based fault diagnosis methods capitalize on the inherent characteristics of industrial systems, providing benefits such as strong interpretability and high diagnostic accuracy. These methods are especially effective in industries like aerospace, where mechanistic models are clearly defined, or for specific components within complex systems that allow for detailed mechanistic insights. However, modern industrial processes frequently display complex traits, including multi-level structures, strong coupling, and uncertainty. These complexities complicate mechanistic analysis and make it challenging to create precise mathematical models. Consequently, the broader application of model-based fault diagnosis methods in large, complex industrial processes is often constrained.

1.2.3 Data-Driven Fault Diagnosis

As illustrated in Fig. 1.4, by gathering extensive training data, techniques such as machine learning and data mining can be employed to extract features related to faults and develop fault classification models, facilitating the predicting, detecting and identifying of faults within the system. Here we analyze three major aspects of

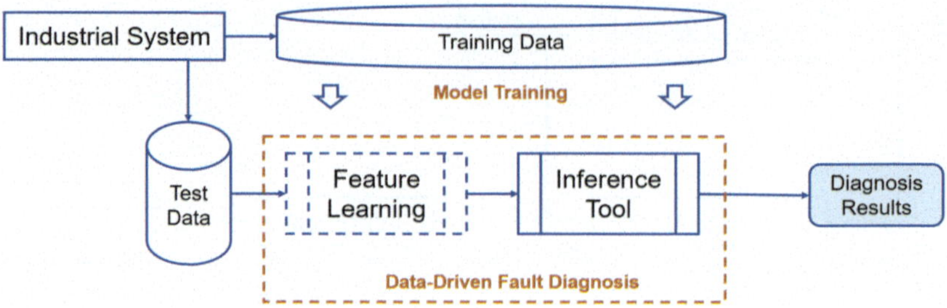

Fig. 1.4 The illustration of data-driven fault diagnosis

data-driven fault diagnosis: signal-based approaches, shallow learning methods, and deep learning methods.

Signal-Based Approaches

Signal-based fault diagnosis methods leverage signals collected from sensors to identify fault types through feature analysis. Common techniques in this domain include time-domain analysis, frequency-domain analysis, and time-frequency analysis [13].

Time-domain analysis focuses on the patterns of signal amplitude changes over time, allowing the examination of temporal features to diagnose faults. In contrast, frequency-domain analysis utilizes tools such as Fourier transforms to explore spectral characteristics, including frequency, amplitude, and phase. For signals exhibiting transient and dynamic features, time-frequency analysis methods, such as Short-Time Fourier Transform, Wavelet Transform, and Hilbert-Huang Transform, are employed [14–16]. These methods enable a comprehensive understanding of signals, which facilitates effective fault diagnosis in various applications [13].

Signal-based fault diagnosis methods focus on analyzing individual signals, making them particularly effective for diagnosing faults in components with relatively straightforward monitoring data, such as bearings and gears. However, in large and complex industrial processes characterized by numerous variables and coupled, nonlinear subsystems, analyzing a single isolated signal does not provide a complete picture of the overall state of the system. Consequently, signal-based fault diagnosis methods are generally unsuitable for application in large and complex industrial processes.

Shallow Model-Based approaches

Shallow model-based fault diagnosis methods, also known as traditional data-driven fault diagnosis approaches, refer to machine learning techniques that consist of only one layer of learning nodes. These methods typically frame fault diagnosis as a regression or class classification problem, where data with similar anomaly characteristics are grouped into the same fault category. Consequently, much of the research focuses on designing feature learners and classifiers to achieve effective fault classification.

Multivariate statistical analysis is one common method for learning fault features. It analyzes the statistical relationships among multiple variables to discern the underlying patterns in the data. Common multivariate statistical methods include Principal Component Analysis and Independent Component Analysis [17]. Matrix decomposition is another prevalent feature learning technique, which involves breaking down a high-dimensional monitoring data matrix into multiple matrix products to achieve dimensionality reduction and feature learning. Notable methods in this category include dictionary learning [18] and non-negative matrix factorization [19]. To achieve reliable diagnostic results, classifiers such as Naive Bayes, Support Vector Machines, K-Nearest Neighbors, and Random Forests are commonly employed as inference tools [20]. These classifiers perform pattern recognition on the input data or features to predict or identify faults. By analyzing the characteristics of the data, they help in accurately determining the presence and type of faults in the system.

Shallow model-based fault diagnosis methods typically require manual design of feature learning techniques and selection of classifiers, necessitating expertise that can introduce subjectivity and limitations to the approach. Additionally, the fitting capability of shallow models is limited, which may constrain their classification performance in more complex fault diagnosis applications.

Deep Learning-Based Approaches

Inspired by deep learning, an increasing number of researchers are applying deep learning models to fault diagnosis [21]. The structure of deep learning models primarily consists of multiple feature extraction layers followed by a decision layer within a neural network. The feature extraction layers progressively learn to capture characteristics of fault data, while the output layer ultimately predicts, detects or identifies the fault type.

Common deep learning methods for fault diagnosis include autoencoders, deep Boltzmann machines, convolutional neural networks, and recurrent neural networks. The advantage of deep learning models lies in their ability to automatically learn features related to faults from large volumes of historical data, eliminating the need to define a mathematical model of the system or manually design and select features. This reduces reliance on human intervention and minimizes subjective errors.

Deep learning exhibits strong fitting capabilities for big data, effectively handling high-dimensional and nonlinear fault data, often achieving high diagnostic accuracy. However, these deep learning-based fault diagnosis methods require high-quality and accurate data, and the multi-layered learning structure can lead to a lack of interpretability in the model's diagnostic decisions.

1.2.4 Brief Summary

Industrial processes draw on knowledge from various disciplines, requiring substantial human, physical, and temporal resources to create a comprehensive expert knowledge base. This endeavor is often hindered by incomplete expert insights and subjective biases. Moreover, the complexity of industrial processes—characterized by numerous components, multiple levels, and intricate subsystem interactions—

makes it challenging to develop accurate analytical models. Consequently, both experience-based and model-based fault diagnosis methods face limitations in practical applications.

Advancements in technologies such as computer networks and advanced sensors have enabled the collection and storage of vast amounts of monitoring data. In era of industrial big data, it is expected that intelligent fault diagnosis will increasingly utilize emerging technologies and theories, including artificial intelligence and data mining. This approach seeks to deliver effective fault diagnosis without relying on precise mathematical models or extensive expert knowledge, thereby ensuring the stable and reliable operation of industrial processes. This trend marks a significant development in the field of intelligent fault diagnosis.

1.3 Testbeds for Process Fault Diagnosis

In this section, we introduce several testbeds specifically designed for industrial process fault diagnosis. These testbeds serve as controlled environments that simulate real-world industrial processes, allowing researchers and practitioners to investigate and evaluate various fault diagnosis methods. By providing a standardized platform, these testbeds facilitate the comparison of different diagnostic techniques and the assessment of their effectiveness in identifying and addressing faults in complex industrial systems.

1.3.1 Tennessee Eastman Process

The Tennessee Eastman process (TEP) [22] serves as a benchmark process for process monitoring and fault diagnosis. It is a complex chemical process simulator based on the real industrial recycle reactor process. TEP is shown in Fig. 1.5. It is composed of reactor, condenser, compressor, stripper and vapor/liquid separator. Besides, it contains a total of 8 components, including 2 products, a by-product, 4 inputs, and an inert, which are denoted by A, B, C, D, E, F, G, and H. The four inflowing streams are derived from distinct production facilities within the plant complex. Notably, there exists substantial holdup for feed stream 3, specifically for component E. Conversely, feed streams 1 and 2, encompassing components A and D, possess relatively less holdup, while feed stream 4, involving components A and C, exhibits a minimal reservoir. In instances where there is limited holdup, adjustments in the feed flowrates are contingent upon product demand variations in the upstream processes responsible for manufacturing those specific components. Consequently, the concern primarily revolves around the variability in the flow rates of three out of the four feed streams, with particular emphasis on stream 4. The initial objective is to minimize flow variability characterized by frequency content within the range of 12–80 h^{-1} for stream 4. As for feed streams 1 and 2 (comprising feeds A and D), safeguarding against variability within the frequency range of 6–16h^{-1} is deemed essential. Subsequently, the variability in the feed rate of component E is

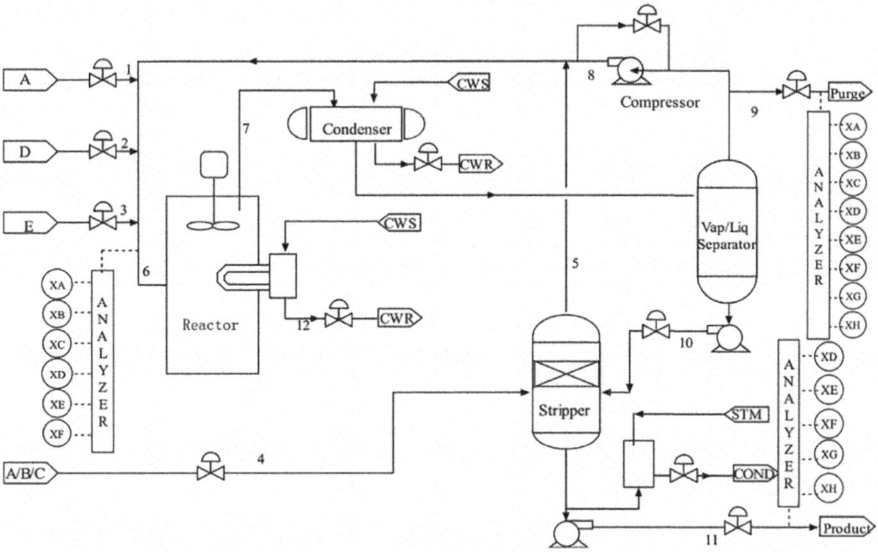

Fig. 1.5 The flowchart of TEP [22]

deemed of lesser significance. However, the occurrence of excessive oscillations or high-frequency ringing (chatter) is undesirable, especially considering the nature of variability exhibited in feed stream 3.

The gaseous reactants are introduced into the reactor, where they undergo catalytic transformations facilitated by a nonvolatile catalyst dissolved in the liquid phase, ultimately yielding liquid products. Within the reactor, an internal cooling bundle is implemented to efficiently dissipate the exothermic heat generated during the chemical reactions. The ensuing vapor-liquid separator serves to segregate non-condensed components. Those components not condensed are recirculated through a centrifugal compressor back into the reactor feed. Condensed components proceed to a product stripping column designed to eliminate residual reactants through a stripping process involving feed stream number 4.

Subsequently, products G and H are discharged from the base of the stripper and directed to a downstream refining section, which is regrettably not elaborated upon in this particular problem statement. The inert and byproduct substances are predominantly purged from the system as vaporous effluents emanating from the vapor-liquid separator.

The TEP dataset is avalabel on a public website.[4] In this dataset, 52 features were generated, which involves 11 control features, 19 composition measurement features and 22 continuous process features. The historical data contained a total of 386 instances of the 20 disturbances IDV(1)–IDV(20), a total of 80 instances of the four setpoint changes SP(1)–SP(4), as well as normal operation. Thus, a total

[4] http://depts.washington.edu/control/LARRY/TE/download.html.

of 24 operating conditions and normal operation were simulated in the historical data. These operating conditions are described in Table 1.1. The available sample sets include training and testing set, containing 22 different simulation operations. Namely, 1 normal and 21 fault operations, which can be seen in Table 1.2. Faults

Table 1.1 Operating conditions of the TE plant with the base control strategy [22]

Op ID	Operating condition	Description	Noise value	Unit Std
N	Normal operation	No disturbances or setpoint changes	0.0024	mol/L
1	IDV(1)	Step in A/C feed ratio, B composition constant	0.71	K
2	IDV(2)	Step in B composition, A/C ratio constant	0.0024	mol/L
3	IDV(3)	Step in D feed temperature (stream 2)	0.71	K
4	IDV(4)	Step in reactor cooling water inlet temperature	0.63	K
5	IDV(5)	Step in condenser cooling water inlet temperature	0.71	L/min
6	IDV(6)	A feed loss (step change in stream 1). Switch pressure controller to purge stream and reduce production rate by 23.8%. Maximum disturbance duration = 72 h	0.71	L/min
7	IDV(7)	C header pressure loss-reduced availability (step change in stream 4)	0.1	L/min
8	IDV(8)	Random variation in A–C feed composition (stream 4)		

(continued)

Table 1.1 (continued)

Op ID	Operating condition	Description	Noise value	Unit Std
9	IDV(9)	Random variation in D feed temperature (stream 2)		
10	IDV(10)	Random variation in C feed temperature (stream 4)		
11	IDV(11)	Random variation in reactor cooling water inlet temperature		
12	IDV(12)	Random variation in condenser cooling water inlet temperature		
13	IDV(13)	Slow drift in reaction kinetics. Maximum disturbance duration = 48 h		
14	IDV(14)	Sticking reactor cooling water valve		
15	IDV(15)	Sticking condenser cooling water valve		
16	IDV(16)	Unknown disturbance. Maximum disturbance duration = 48 h		
17	IDV(17)	Unknown disturbance. Maximum disturbance duration = 48 h		
18	IDV(18)	Unknown disturbance. Maximum disturbance duration = 12 h		

(continued)

Table 1.1 (continued)

Op ID	Operating condition	Description	Noise value	Unit Std
19	IDV(19)	Unknown disturbance		
20	IDV(20)	Unknown disturbance. Maximum disturbance duration = 8 h		
21	SP(1)	Production rate change (step down 15%)		
22	SP(2)	Product mix change (50/50 to 40/60)		
23	SP(3)	Reactor operating pressure change (step down by 60 kPa)		
24	SP(4)	Purge gas: Component B change (step up 2%)		

1–7 are deterministic faults whereas Faults 8–12 are stochastic faults. The data are generated at a sampling interval of 3 min. There are 500 normal samples for normal operation and 480 fault samples for fault operations in training set. For testing set, 960 samples are generated in which the first 160 samples are normal and the remaining ones are abnormal for each fault.

1.3.2 Multi-phase Flow Process

The industrial Multi-phase flow (MPF) facility[5] is a real-world dataset and has been extensively used in the fault diagnosis field [23]. The MPF process, designed by Cranfield University, is a complex pressurized system working under varying operation conditions. It is simply described in Fig. 1.6. The experimental setup encompasses pipelines with diverse bore sizes and geometries, culminating in a gas and liquid two-phase separator situated atop a 10.5 m elevated platform, featuring a diameter of 0.5 m and a height of 1.2 m. The system is capable of receiving single-phase inputs

[5] http://www.mathworks.com/matlabcentral/fileexchange/50938-a-benchmark-case-for-statistical-process-monitoring-cranfield-multiphase-flow-facility.

Table 1.2 Process faults in the TEP [22]

No.	Description	Type	Sample number
1	A/C feed ratio, B composition constant (stream 4)	Step	Density top riser (kg/m^3)
2	B composition, A/C ratio constant (stream 4)	Step	Density top separator output (kg/m^3)
3	D feed temperature (stream 2)	Step	Density water input (kg/m^3)
4	Reactor cooling water inlet temperature	Step	Temperature top riser (°C)
5	Condenser cooling water inlet temperature	Step	Temperature top separator output (°C)
6	A feed loss (stream 1)	Step	Temperature water input (°C)
7	C header pressure loss—reduced availability (stream 4)	Step	Gas-liquid three-phase separator (kg/s)
8	A, B, C feed composition (stream 4)	Random variation	VC501 valve position (kg/s)
9	D feed temperature (stream 2)	Random variation	VC302 valve position (kg/s)
10	C feed temperature (stream 4)	Random variation	VC101 valve position (kg/s)
11	Reactor cooling water inlet temperature	Random variation	Pump current (A)
12	Condenser cooling water inlet temperature	Random variation	
13	Reaction kinetics	Slow drift	
14	Reactor cooling water valve	Sticking	
15	Condenser cooling water valve	Sticking	
16	Unknown	Unknown	
17	Unknown	Unknown	
18	Unknown	Unknown	
19	Unknown	Unknown	
20	Unknown	Unknown	
21	The valve for stream 4 was fixed at the steady-state position	Constant position	

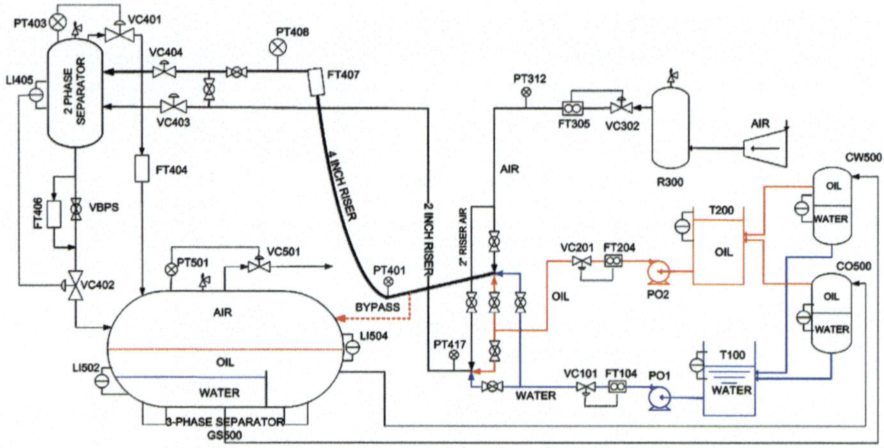

Fig. 1.6 The flowchart of MPF process [23]

of air, water, and oil, or a combination thereof, at specified rates. Subsequently, the fluid mixtures undergo separation within an 11 m^3 horizontal three-phase separator located at ground level (denoted as GS500). The separated air is released into the atmosphere, while the emulsions of oil and water undergo further separation in dedicated coalescers, denoted as CW500 and CO500, each boasting a capacity of approximately 1.5 m^3. These fluids then return to their respective storage tanks, T200 and T100, both with a capacity of around 12.5 m^3.

The air supply is facilitated by a union of two compressors capable of delivering a flow rate of up to 0.4 m^3/s at 0.7 MPa. The compressed air is received in an 8 m^3 vessel (labeled as R300) to mitigate pressure fluctuations. Following this, the air undergoes filtration to remove droplets and particles, followed by cooling. The air flow is monitored by the air flow meter FT305, and its flow is controlled by the pneumatic valve VC302.

Water and oil are stored in tanks T100 and T200, respectively, and are independently supplied through multistage Grundfos CR90-5 pumps (PO1 and PO2). Each pump can provide up to 0.0278 m^3/s at 1 MPa, with their rotational speed controlled by frequency variable inverters. The water flow rate is measured by FT104, and the oil flow rate is monitored by FT204. Pneumatic valves VC101 and VC201 control the flow rates of water and oil, respectively.

Post-mixing, the fluid streams can traverse either a 4" diameter flow loop, featuring a 55 m long and 21 downward-inclined pipeline leading to a 10.5 m high catenary riser, or a 2" flow loop, comprising a 40 m long horizontal pipeline connected to a 10.5 m long vertical riser. Both flowlines are linked to the two-phase separator, with the ability to be isolated from each other via manual valve manifolds at both ends of the flowlines. Throughout all conducted experiments in this benchmark case, the 4" line was exclusively utilized, except for the investigation of a specific fault scenario (fault case 6), where the 2" line was also engaged. Notably, the rig incorporates an alternative 4" line labeled as "BYPASS," depicted in dashed red in Fig. 1.6, allowing direct flow from the mixing point to the three-phase separator, circumventing the riser and the top separator.

In the experimental setup, pressure measurements are obtained from various strategic locations within the system. Specifically, sensors are deployed to monitor the air supply line pressure before the mixing point (PT312), the pressure along the 4" line at the bottom of the riser (PT401), the pressure at the top of the riser (PT408), as well as pressures within the top separator (PT403) and the three-phase separator (PT501). The flow rate in the 4" line at the top of the riser is quantified by FT407, which concurrently provides measurements of density and temperature. Furthermore, FT406 facilitates the determination of mass flow rate, density, and temperature at the bottom outlet of the top separator.

The liquid level inside the top separator is gauged by LI405 and regulated by VC402. Control of the pressure within the three-phase separator is managed by VC501, with LI502 and LI504 serving to measure the levels between different phases.

Table 1.3 Variables description for multi-phase flow process [23]

No.	Description	No.	Description
1	Conveying air pressure (MPa)	13	Density top riser (kg/m^3)
2	Pressure at the bottom of the riser (MPa)	14	Density top separator output (kg/m^3)
3	Pressure at top of the riser (MPa)	15	Density water input (kg/m^3)
4	Pressure at top separator (MPa)	16	Temperature top riser (°C)
5	Pressure in three-phase separator (MPa)	17	Temperature top separator output (°C)
6	Diff. pressure (PT401-PT408) (MPa)	18	Temperature water input (°C)
7	Differential pressure above VC404 (MPa)	19	Gas-liquid three-phase separator (kg/s)
8	Flow input air (m^3/s)	20	VC501 valve position (kg/s)
9	Inlet water flow (kg/s)	21	VC302 valve position (kg/s)
10	Flow top riser (kg/s)	22	VC101 valve position (kg/s)
11	Horizontal top separator (m)	23	Pump current (A)
12	Flow top separator output (kg/s)		

Additionally, PT417 monitors the pressure in the 2" line at the bottom of the riser. For the sake of simplicity, valves and sensors responsible for controlling the level between phases in the coalescers have not been depicted in Fig. 1.6. The testing area is composed of a gas-liquid two-phase separator and pipelines with various geometries and bore sizes.

In the MPF dataset, 24 process variables were used alongside two process inputs, specifically the set points for water and air flow rates, as detailed in Table 1.3. All data were recorded at a sampling frequency of 1 Hz. The dataset includes six fault conditions under both steady-state and varying scenarios, which encompass different flow rate set points, air line blockage, water line blockage, top separator input blockage, open direct bypass, slugging conditions, and pressurization of the #2 line.

1.3.3 Continuous Stirred-Tank Reactor

The Continuous Stirred Tank Reactor (CSTR) [24] is characterized as a dynamic nonlinear system, exemplified by the feedback control system illustrated in Fig. 1.7. The CSTR receives a flow of solvent and reactant A, initiating a first-order irreversible reaction A to B within the reactor, ultimately producing B as the output stream. To ensure the stability of the reactor system, a closed-loop control system is implemented for the regulation of the reactor temperature. Within the simulation framework, 8 process variables are meticulously gathered to facilitate process fault diagnosis, each subjected to the introduction of Gaussian noises. The specifics of the eight measurement variables, along with their respective noise standard deviations, are comprehensively delineated in Table 1.4.

Under normal operating conditions, a set of 500 samples is generated to form the training set. To emulate real-world scenarios and assess system robustness, three distinct failure modes are introduced in the experiment. In each fault scenario, a total of 800 samples are generated, with the fault manifesting in the 301st sample. The three delineated fault types encompass alterations in operating conditions, variations in process parameters, and discrepancies in sensor readings.

1.3.4 Fed-Batch Fermentation Penicillin Process

The Fed-batch fermentation penicillin process[6] (FBFP) [25], as shown in Fig. 1.8, stands as a prominent industrial benchmark process, extensively employed for the validation of process fault diagnosis methodologies. FBFP emulates a nonlinear industrial process encompassing the cellular life cycle from growth to autolysis. The intricacy of the FBFP process arises from its high sensitivity to various critical

[6] http://simulator.iit.edu/web/pensim/simul.html.

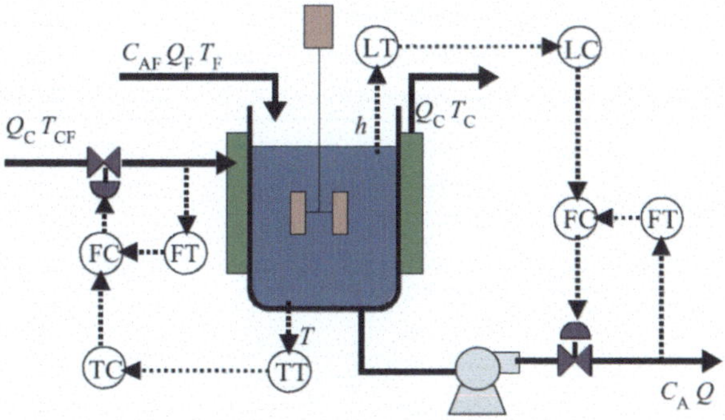

Fig. 1.7 The flowchart of CSTR system [24]

Table 1.4 Measurement variables and noise standard deviation of CSTR system [24]

Variable name	Variable description	Normal	Noise value	Unit Std
C_A	Concentration of reactant A in reactor	0.0372	0.0024	mol/L
T	Reactor temperature	402.35	0.71	K
C_{AF}	Concentration of reactant A in reactor feed stream	1.0	0.0024	mol/L
T_F	Reactor feed temperature	320.0	0.71	K
T_{CF}	Coolant inlet temperature	300	0.63	K
Q	Reactor outlet flow rate	100	0.71	L/min
Q_F	Reactor inlet flow rate	100	0.71	L/min
Q_C	Coolant flow rate	15	0.1	L/min

process variables, notably including temperature, pH, substrate concentration, and other influential factors. The datasets can be simulated from PenSim v2.0, including 11 process variables, as listed in Table 1.5, which includes 11 process variables. The whole duration of each batch was 400 h, including 45 h of preculture phase and 355 h of fed-batch stage, and the sampling interval is 1 h.

FBFP is a prototypical dynamic, non-linear, time-varying, and multi-stage batch process with distinct phases. Broadly categorized, the process unfolds in three sequential stages: the cell growth stage, the penicillin synthesis stage, and the cell autolysis stage. During the initial cell growth phase, nutrients from the substrate are consumed, leading to continuous synthesis of new cells. Subsequently, the penicillin

Table 1.5 Process variables of FBFP [25]

Number	Process variable	Unit	Noise value	Unit Std
1	Aeration rate	L/h	0.0024	mol/L
2	Agitator power	W	0.71	K
3	Substrate feed temperature	K	0.0024	mol/L
4	Substrate concentration	g/L	0.71	K
5	Biomass concentration	g/L	0.63	K
6	Penicillin concentration	g/L	0.71	L/min
7	Culture volume	L	0.71	L/min
8	Carbon dioxide concentration	g/L	0.1	L/min
9	pH			
10	Fermenter temperature	K		
11	Cooling water flow rate	L/h		

synthesis stage commences, characterized by peak productivity until the synthesis capacity of penicillin diminishes. The concluding stage is cell autolysis, marked by an increase in pH within the fermentation broth and a gradual decline in penicillin synthesis capability. The pH was rigorously maintained at a constant level of 5.0 to emulate the observed dynamics of penicillin production, employing either an on/off or a proportional-integral-derivative (PID) controller. pH regulation was achieved through the judicious addition of highly concentrated (3 M) acid or base solution as deemed necessary. Dual PID controllers were employed to modulate the control values of acid and base. The tuning of these PID controllers was meticulously performed to optimize performance within a predefined range of initial conditions, considered representative of normal operational parameters. The temperature of the culture medium was kept constant at 25 °C. A PID controller was designed to control the temperature of the culture by manipulating heating/cooling water flow rate.

1.3.5 Hot Steel Rolling Process

The hot steel rolling process [26], depicted in Fig. 1.9, involves five key steps: billet preparation, heating, rough rolling, finishing rolling, and cooling. This dataset focuses on a critical component of this process: the conveyor rollers. Steel conveyor rollers are essential for transferring steel between different stages of production.

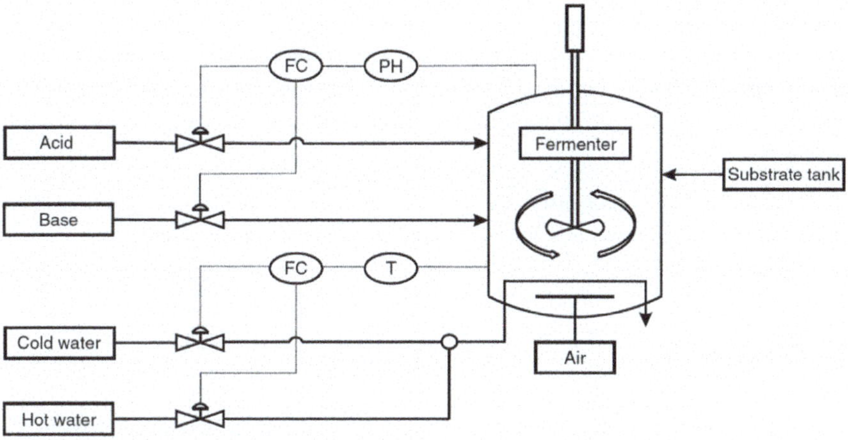

Fig. 1.8 The flowchart of FBFP system [25]

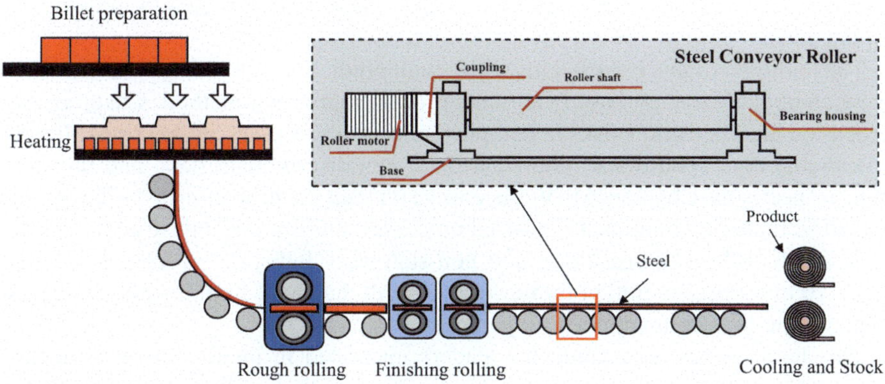

Fig. 1.9 The process of steel hot rolling [26]

The data were collected from a real hot steel rolling industry and include information from a total of 294 steel conveyor rollers (SCR). The operating status of each roller was monitored using its motor current signal, which was converted into digital data for analysis. The dataset spans six months, from January 1, 2023, to June 6, 2023, and encompasses five categories of fault data, as detailed below:

- **Shaft Swing**: A fault characterized by irregular lateral movement or wobbling of the roller's shaft.
- **Shaft Stuck**: A condition where the roller's shaft becomes immobilized or experiences significant resistance to its intended rotational movement.
- **Over-Current**: A situation where the electrical current supplied to the roller motor exceeds its designed or safe operating limits.

- **Squeak**: An audible, high-pitched noise produced during operation, typically caused by friction or inadequate lubrication between moving components.
- **Base Deformation**: A condition where the base experiences a physical alteration or distortion, resulting from factors such as excessive weight, impact, or structural weakness.

These five faults occur naturally during daily operations, leading to temporary suspensions of the rolling process for necessary inspections and repairs.

1.4 Preface

This book provides a comprehensive introduction to big data-driven intelligent fault diagnosis strategies for industrial processes, covering fault prediction, detection, and identification within the broader context of fault diagnosis. All methodologies discussed are validated using either public datasets or practical data from industrial applications.

The reminder of this book begins with a fault prediction method in Chap. 2, which using an update-strategy-based Gaussian process regression method. Chapters 3 to 5 focus on fault detection tasks in complex industrial processes, addressing different undesirable data conditions: redundant data, non-discriminant data, and dynamic Data, respectively. Chapters 6 to 9 concentrate on fault identification tasks for various scenarios, including nonlinear industrial processes, multi-mode industrial processes, zero-sample industrial processes, and non-stationary industrial processes. Finally, Chap. 10 introduces a unified framework for fault diagnosis in industrial processes, emphasizing its incremental learning capabilities.

This book serves as a guide for readers interested in intelligent fault diagnosis of industrial processes. It presents fundamental concepts alongside cutting-edge research and state-of-the-art algorithms. We anticipate that readers from diverse research areas related to industry would find valuable insights in this book.

References

1. Isermann R, Balle P (1997) Trends in the application of model-based fault detection and diagnosis of technical processes. Control Eng Pract 5(5):709–719
2. Chiang LH (2000) Fault detection and diagnosis in industrial systems. Springer Science & Business Media (2000)
3. Angeli C (2008) Online expert systems for fault diagnosis in technical processes. Expert Syst 25(2):115–132
4. Lee W-S, Grosh DL, Tillman FA, Lie CH (1985) Fault tree analysis, methods, and applications: a review. IEEE Trans Relia 34(3):194–203
5. Qian Yu, Li X, Jiang Y, Wen Y (2003) An expert system for real-time fault diagnosis of complex chemical processes. Expert Syst Appl 24(4):425–432

6. Lavasani SM, Zendegani A, Celik M (2015) An extension to fuzzy fault tree analysis (ffta) application in petrochemical process industry. Process Safety Environ Protect 93:75–88
7. Vedam H, Venkatasubramanian V (1999) Pca-sdg based process monitoring and fault diagnosis. Control Eng Pract 7(7):903–917
8. Wu J-D, Bai MR, Su F-C, Huang C-W (2009) An expert system for the diagnosis of faults in rotating machinery using adaptive order-tracking algorithm. Expert Syst Appl 36(3):5424–5431
9. Gao Z, Cecati C, Ding SX (2015) A survey of fault diagnosis and fault-tolerant techniques—part i: fault diagnosis with model-based and signal-based approaches. IEEE Trans Indus Electron 62(6):3757–3767
10. Isermann R (1993) Fault diagnosis of machines via parameter estimation and knowledge processing–tutorial paper. Automatica 29(4):815–835
11. Xin W, Liu Y (2017) Leakage detection for hydraulic igv system in gas turbine compressor with recursive ridge regression estimation. J Mecha Sci Technol 31:4551–4556
12. Patton RJ, Chen J (1991) A review of parity space approaches to fault diagnosis. IFAC Proc Vol 24(6):65–81
13. Lei Y, Yang B, Jiang X, Jia F, Li N, Nandi AK (2020) Applications of machine learning to machine fault diagnosis: a review and roadmap. Mech Syst Signal Process 138:106587
14. Wang L-H, Zhao X-P, Jia-Xin W, Xie Y-Y, Zhang Y-H (2017) Motor fault diagnosis based on short-time fourier transform and convolutional neural network. Chinese J Mech Eng 30:1357–1368
15. Yan R, Gao RX, Chen X (2014) Wavelets for fault diagnosis of rotary machines: a review with applications. Signal Process 96:1–15
16. Peng ZK, Tse Peter W, Chu FL (2005) A comparison study of improved hilbert–huang transform and wavelet transform: Application to fault diagnosis for rolling bearing. Mech Syst Signal Process 19(5):974–988
17. Kruger U, Xie L (2012) Statistical monitoring of complex multivatiate processes: with applications in industrial process control. Wiley
18. Li Y, Chai Y, Yin H (2021) Autoencoder embedded dictionary learning for nonlinear industrial process fault diagnosis. J Process Control 101:24–34
19. Zhou H, Yin H, Chai Y (2023) Multi-grained mode partition and robust fault diagnosis for multimode industrial processes. Reliab Eng Syst Safety 231:109011
20. Rauber TW, do Nascimento EM, Wandekokem ED, Varejão FM, Herout A (2010) Pattern recognition based fault diagnosis in industrial processes: review and application. Pattern Recognition Recent Adv, 483–508
21. Jianbo Yu, Zhang Y (2023) Challenges and opportunities of deep learning-based process fault detection and diagnosis: a review. Neural Comput Appl 35(1):211–252
22. Downs JJ, Vogel EF (1993) A plant-wide industrial process control problem. Comput Chem Eng 17(3):245–255
23. Ruiz-Cárcel C, Cao Y, Mba D, Lao L, Samuel RT (2015) Statistical process monitoring of a multiphase flow facility. Control Eng Pract 42:74–88
24. Wayne Bequette B (2003) Process control: modeling, design, and simulation. Prentice Hall Professional
25. Birol G, Ündey C, Cinar A (2002) A modular simulation package for fed-batch fermentation: penicillin production. Comput Chem Eng 26(11):1553–1565
26. Zhao D, Yin H, Zhou H, Cai L, Qin Y (2024) A zero-sample fault diagnosis method based on transfer learning. IEEE Trans Indus Inform

Chapter 2
Update-Strategy-Based Gaussian Processes Regression for Processes Fault Prediction with Incremental Data

Health state prediction and fault time prediction are pivotal in fault prognostics. Traditional approaches often treat these tasks separately, overlooking the time-varying dynamics of system operations, which compromises efficiency and accuracy. To address this, we propose a Gaussian Process Regression (GPR) method with an adaptive update strategy tailored for aeroengine prognostics. This approach continuously refines model parameters and predictive distributions using new data. By integrating variable selection and multivariable fusion, it identifies key health-related variables to construct a comprehensive health index, enabling simultaneous health state and fault time prediction. The method employs single-point and multipoint update strategies within a unified GPR framework, effectively incorporating new sample data for dual-task predictions. Experimental results on the C-MAPSS dataset demonstrate that the proposed method outperforms state-of-the-art techniques.

2.1 Motivation

Considering the complex design and essential operational function of aeroengines, which are expensive machines, prognostics and health management (PHM) have become essential in this sector. Recently, motivated by the need for reliability, safety, and reduced maintenance costs, traditional maintenance approachee transitioned from scheduled and preventive methods to predictive and condition-based strategies [1]. Health state prediction and fault time prediction represent two vital but challenging aspects of these maintenance practices [2]. Fault refers to a condition where the system is unable to perform its specified function due to the failure of certain components. By analyzing collected historical data, the degradation process is characterized using a constructed health index (HI), enabling accurate assessment of the degradation level and its evolutionary trend through health state prediction. Fault time prediction, on the other hand, estimates the time of fault occurrence in advance, allowing for adjustments in maintenance plans. This approach maximizes system

H. Yin et al., *Data-Driven Fault Diagnosis for Complex Industrial Processes*, Engineering Applications of Computational Methods 22,
https://doi.org/10.1007/978-981-96-3153-7_2

uptime while minimizing maintenance costs. Consequently, accurate fault prediction provides valuable insights to prevent unexpected machine shutdowns, thereby ensuring system dependability and safety, as well as reducing maintenance expenses.

In recent years, the advent of sensor technology has led to the widespread use of various sensors to monitor the health state of systems or units. The large volumes of collected sensor data have prompted the evolution of traditional fault prediction methods into data-driven approaches. Despite the extensive application of these machine learning methods, fault prediction remains fundamentally probabilistic. The results produced by these methods typically represent one-step-ahead predictions without uncertainty representation [3], thereby underscoring the efficacy of probability-based methods as viable alternatives. For example, Xu et al. [3] introduced a health prediction method incorporating health precaution levels for sensor systems, and Chen et al. [4] implemented aeroengine Remaining Useful Life (RUL) prediction based on particle filters. The Hidden Markov Model (HMM), a probability-based model, has been applied to fault diagnosis and prognosis, exemplified by Soleimani et al. [5], who combined Bayesian networks and HMM for fault detection, prediction, and isolation. In the realm of fault prognosis, considerable attention has been devoted to health state prediction and fault time prediction within PHM. Duan et al. [6] proposed a fault time prediction framework for hidden-state systems using a multi-attribute Bayesian control chart. Recently, deep learning methods have seen extensive use in this context. For instance, Li et al. [7] developed a nonlinear neural data fusion method for HI construction, and Zhang et al. [8] utilized a dual-task network scheme for health assessment and RUL prediction in aeroengines.

However, most state-of-the-art techniques perform HI and fault time prediction separately and hierarchically, without considering the time-varying dynamics of the degradation process. Specifically, these methods often overlook the mapping between dual tasks, resulting in the absence of a unified framework for dual-task prediction. Moreover, due to the time variability of the degradation process within systems or units, the accuracy of prediction models significantly declines if they continue to be executed with fixed parameters after new degradation data becomes available. Based on the identified issues, this chapter proposes a novel fault prediction methodology utilizing an update-strategy-based GPR for health state assessment and fault time prediction. Given the extensive variability in HI across different systems or machines, an unsupervised HI construction method is introduced following the selection of variables that contribute most significantly. This approach obviates the necessity for a preset fault threshold. Subsequently, two update strategies for GPR models—single-point and multipoint—are developed to enhance fault prediction accuracy.

The contributions of this chapter are highlighted as follows:

- An unsupervised HI construction method, integrating a variable selection algorithm and multivariable fusion technology, is developed to accurately reflect the health state of the system. This method autonomously sets the HI at the fault status to 0, thereby addressing the uncertainty associated with fault thresholds.
- Two fault prediction models employing GPR with update strategies are proposed to deliver more precise fault prediction information. The update-strategy-based GPR

allows for dynamic parameter updating, thereby improving prediction accuracy and efficiency.

- A dual-task prognosis scheme is implemented to enhance the performance of HI and fault time prediction within a unified framework. The shared HI-based approach facilitates simultaneous HI and fault time prediction, achieving a cost-effective and straightforward mapping process for dual tasks.

2.2 Methodology

2.2.1 An Unsupervised HI Construction Method

While full-life monitoring signals (variables) of a system contain vital information that spans the entire range from normal to fault states, not all variables effectively represent the system's condition. Specifically, some health state-related variables demonstrate a strong health state with a high signal-to-noise ratio, whereas others may not [9]. Furthermore, data fusion that overlooks the system's health state can lead to an independent HI. As a result, these raw signals undergo analysis and variable selection. Consequently, to more accurately represent the system's health state, the HI is constructed.

In this context, the z-score standardization technique, as indicated in Eq. (2.1), is utilized to mitigate the influence of outliers in the raw signals.

$$S_{sta} = \frac{s - \mu(s)}{\delta(s)} \tag{2.1}$$

where $\mu(s)$ and $\delta(s)$ are the mean and standard deviation of the raw signal s, respectively. The mean and standard deviation of the standardized data S_{sta} are 0 and 1.

The objective of variable selection is to identify the most informative variables pertinent to health states, which significantly influence the system state. In this context, the Kullback-Leibler Divergence (KLD) is utilized as a criterion to measure the divergence of a variable between normal and abnormal states. A variable with a high KLD value indicates a substantial contribution to reflecting the system's health state. Conversely, variables with low KLD values may not effectively indicate the correlation between system variables and the health state. Additionally, measurement noise can also result in low KLD values. Therefore, the cumulative contribution rate (CCR) is employed to address this issue.

$$CCR_p = \frac{\sum_{i=1}^{p} KLD_i}{\sum_{i=1}^{q} KLD_i} \times 100\% \tag{2.2}$$

where KLD_i denotes the ith largest KLD values. The parameters p and q are respectively the number of the state-relevant variables and raw variables. Given a CCR limit, the parameter p can be determined.

While the state-relevant variable can be obtained using KLD, the noise in the raw variable can weaken the correlation between the constructed HI and the actual health state.

Consequently, data denoising is essential prior to the construction of the HI. Moving average filtering, a widely used data denoising technique, is effective in mitigating the impact of random disturbances and significant signal fluctuations. This method can be mathematically defined as:

$$v_{i,n} = \frac{1}{L} \sum_{k=-(L-1)/2}^{(L-1)/2} S_{sta,n-k}, i = 1, 2, \ldots, p \tag{2.3}$$

where L is an odd sliding window width, and $v_{i,n}$ is nth filtered output of the ith state-relevant variable. $S_{sta,n-k}$ is the $n-k$th point in standardized data.

Since only partial information about the health state is captured for each variable derived from the raw sensor data, multivariable fusion can provide more accurate and robust inferences [9]. To characterize the system's health state, multiple state-relevant variables are combined to form a single Health Index HI.

For the set of health state-relevant variables $V = [v_1, v_2, \ldots, v_p]$, a new HI is developed through a linear combination, which can be expressed as follows:

$$HI = w_1 v_1 + w_2 v_2 + \cdots + w_1 v_1 = WV^T \tag{2.4}$$

where W represents the weight matrix. In contrast to other methods, the weight matrix W is derived from KLD results. A higher weight value signifies greater relevance of the corresponding variable to the system's health state.

$$w_i = sgn(M_{v_i}) \frac{KLD_i}{\sum_{k=1}^{p}}, i = 1, 2, \ldots, p \tag{2.5}$$

where $sgn(\cdot)$ is a symbolic function. M_{v_i} represents the monotonicity of the variable v_i. If v_i monotonically decrease, $sgn(M_{v_i})$ takes 1, and vice versa.

The developed HI at each cycle can be represented as $HI = [h_1, h_2, \ldots, h_t]$. In this context, the HI varies from 1 to 0 as the system transitions from a healthy state to a fault state, similar to [3]. Specifically, $HI = 0$ indicates that a fault has occurred. Therefore, the HI must be normalized to fall within the range of 0 to 1 through a normalization procedure:

$$\hat{h}_i = \frac{h_i - h_{min}}{h_{max} - h_{min}}, i = 1, 2, \ldots, t \tag{2.6}$$

where h_{max} and h_{min} are respectively the maximum and minimum of HI. $\hat{H}I = [\hat{h}_i, \hat{h}_2, \ldots, \hat{h}_t]$ is the normalized HI matrix, which shows the trajectory of the system health state.

In the current methodology, the HI curve is generated in an unsupervised manner using state-dependent variables, without incorporating any domain-specific knowledge or manually labeled HI values. To assess the relationship between the constructed HI and the system's health state, three criteria are introduced. The monotonicity (Mon) criterion reflects the trend of the HI, the correlation (Cor) criterion uncovers the nonlinear relationship between the HI and running time, and the distance (Dis) metric evaluates the exponential properties. A higher score for both Mon and Cor, coupled with a lower score for Dis, indicates improved performance of the constructed HI. These criteria are defined as follows:

$$Mon = \frac{N_{d\hat{H}I>0} - N_{d\hat{H}I<0}|}{t-1} \tag{2.7}$$

$$Cor = 1 - \frac{6\sum_{k=1}^{t}(rg(\hat{h}_k)_r g(T_k))^2}{t(t^2-1)} \tag{2.8}$$

$$Dis(\hat{H}I, Z) = \sqrt{\sum_{k=1}^{t}(\hat{h}_K - z_k)^2} \tag{2.9}$$

where $d\hat{H}I - \hat{h} - \hat{h}_{t-1}$, $N_{d\hat{H}I>0}$ and $N_{d\hat{H}I<0}$ indicate the number of $d\hat{H}I$ greater and less than 0, respectively. The parameter t is the length of the constructed HI. $rg(\hat{h}_k)$ and $rg(T_k)$ are respectively the ranks of the kth observation point in the constructed HI matrix $\hat{H}I$ and running time matrix T. $Z = [z_1, z_2, \ldots, z_t]$ is the corresponding empirical degradation curve of $\hat{H}I$. It is obtained by fitting a two-term exponential function, which has widespread acceptance in degradation trends (HI curves) [10, 11].

2.2.2 *Single-Point Update Strategy*

To create an accurate and robust prediction framework, it is essential to carefully select health-dependent variables to optimize parameters. While the chosen subset of variables includes operational conditions and the most relevant information, the system's degradation process is dynamic and time-varying. As a result, after fault prediction is conducted, new degradation data may become available during equipment operation, leading to a gradual decline in the prediction accuracy of models based on fixed parameters. An update mechanism is a commonly used strategy for continuously obtaining optimal model parameters.

In the prediction process, the single-point update strategy focuses on updating the model parameters only when new sample data becomes available for each prediction. For the regression problem with $\varepsilon_m \sim N(0, \sigma_m^2)$, the prior distribution of outputs y_m with m historical sample points is expressed as:

$$y_m = N(\mu_m, k_{mm} + \sigma_m^2 I_m) \tag{2.10}$$

where $k_{mm} = \begin{bmatrix} k(x_1, x_1) & \cdots & k(x_1, x_m) \\ \vdots & \ddots & \vdots \\ k(x_m, x_1) & \cdots & k(x_m, x_m) \end{bmatrix}$

To update the prediction model, the input and output of a new sample (x_+, y_+) must be incorporated into the relevant historical datasets. When a new sample point is collected, the single-point update mechanism is activated. According to (7), the joint prior Gaussian distribution for the new sample point can be defined as:

$$\begin{bmatrix} \begin{bmatrix} y_m \\ y_+ \end{bmatrix} \\ y^* \end{bmatrix} \sim N\left(\begin{bmatrix} \begin{bmatrix} \mu_m \\ x_+ \end{bmatrix} \\ \mu_* \end{bmatrix}, \begin{bmatrix} \begin{bmatrix} k_{mm} + \sigma_m^2 I_m & k_{m+} \\ k_{+m} & k_{++} \end{bmatrix} & \begin{bmatrix} k_{m*} \\ k_{+*} \end{bmatrix} \\ \begin{bmatrix} k_{*m} & k_{*+} \end{bmatrix} & k_{**} \end{bmatrix} \right) \tag{2.11}$$

Equation (2.11) can be simplified as follows:

$$\begin{bmatrix} y_+^m \\ y^* \end{bmatrix} \sim N\left(\begin{bmatrix} \mu_+^m \\ \mu_* \end{bmatrix}, \begin{bmatrix} k_+^m & k_*^T \\ k_* & k_{**} \end{bmatrix} \right) \tag{2.12}$$

where $y_+^m = [y_m, y_+]^T$, $\mu_+^m = [\mu_m, x_+]^T$,$k_* = [k_{*m}\ k_{*+}]$, $k_{+m} = \begin{bmatrix} k_{mm} + \sigma_m^2 I_m & k_{m+} \\ k_{+m} & k_{++} \end{bmatrix}$.

The newly formulated predictive distribution, which incorporates Gaussian noise, adheres to the properties of a Gaussian distribution $N(\mu, \sum)$.

$$\mu = k_*(k_+^m)^{-1}(y_+^m - \mu_+^m) + \mu_* \tag{2.13}$$

$$\sum = k_{**} - k_*(k_+^m)^{-1} k_*^T \tag{2.14}$$

It is clear to see that the time complexity of new predictive distribution under single-point update strategy is $O\left((m+1)^3\right)$ due to the inverse matrix $(k_+^m)^{-1}$. In fact, the matrices $k_{mm} + \sigma_m^2 I_m$ and k_{++} are available. Hence, the new predictive distribution $N(\mu, \sum)$ under single-point update strategy can be further optimized.

The inverse matrix $(k_+^m)^{-1}$ can be induced as follows:

$$\begin{aligned} (k_+^m)^{-1} &= \begin{bmatrix} K + K^{-1} k_{m+} \Lambda k_{+m} K^{-1} & -k_{mm} k_{m+} \Lambda \\ -\Lambda k_{+m} k_{mm} & \Lambda \end{bmatrix} \\ &= \begin{bmatrix} K & 0 \\ 0 & 0 \end{bmatrix} + \begin{bmatrix} K^{-1} k_{m+} \\ -1 \end{bmatrix} \Lambda [k_{+m} K^{-1} \ -1] \end{aligned} \tag{2.15}$$

where $\Lambda = [k_{++} - k_{+m} K^{-1} k_{m+}]^{-1}$, $K + k_{mm} + \delta_m^2 I_m$.

Because Λ is a scalar, the time complexity depends on the matrix K_{-1}. Thus, the time complexity of optimized predictive distribution is $O(m^3)$, which is less than $O\left((m+1)^3\right)$.

2.2.3 Multi-point Update Strategy

In contrast to the single-point update strategy, the multipoint update strategy focuses on updating model parameters using a sample set that includes multiple new data points. While the prediction results from repeatedly applying the single-point update may approximate those from the multipoint update, the latter is crucial for streamlining the computational process. Additionally, during the early stages of degradation, the single-point update strategy proves to be suboptimal. For the newly collected sample set (x_n, y_n) with n data points, it is assumed that the output $y_n \sim N(\mu_m, k_{mm} + \sigma_n^2 I_n)$. The joint distribution between the historical output y_n and the new sample output y_n can be expressed as follows:

$$\begin{bmatrix} y_m \\ y_n \end{bmatrix} \sim N\left(\begin{bmatrix} \mu_m \\ \mu_n \end{bmatrix}, \begin{bmatrix} k_{mm} + \delta_m^2 I_m & k_{mn} \\ k_{nm} & k_{nn} + \delta_n^2 I_n \end{bmatrix}\right) \tag{2.16}$$

For the multipoint update task, the conditional probability $p(y_m | x_m, x_n, y_n)$ needs to be deduced. Thus, Eq. (2.16) is rewritten as follows:

$$\begin{bmatrix} y_n \\ y_m \end{bmatrix} \sim N\left(\begin{bmatrix} \mu_n \\ \mu_m \end{bmatrix}, \begin{bmatrix} k_{nn} + \delta_n^2 I_n & k_{nm} \\ k_{mn} & k_{mm} + \delta_m^2 I_m \end{bmatrix}\right) \tag{2.17}$$

Based on Bayesian theory and the properties of Gaussian distribution, the distribution $y_m^n = y_m | y_n$ of adding a new sample set also confirms to Gaussian distribution. Namely,

$$y_m^n \sim p(y_m | x_m, x_n, y_n) = p(y_m | y_n) = N(\mu_m^n, \Sigma_m^n) \tag{2.18}$$

where

$$\mu_m^n = k_{mn}(k_{nn} + \delta_n^2 I_n)^{-1}(y_n - \mu_n) + \mu_m \tag{2.19}$$

$$\Sigma_m^n = k_{mm} + \delta_m^2 I_m - k_{mn}(k_{nn} + \delta_n^2 I_n)^{-1} k_{nm} \tag{2.20}$$

At this stage, the posterior distribution after incorporating the new sample set is obtained. The joint distribution related to a new input set x^* can be inferred as follows:

$$\begin{bmatrix} y_m^n \\ y^* \end{bmatrix} \sim N\left(\begin{bmatrix} \mu_m^n \\ \mu_* \end{bmatrix}, \begin{bmatrix} k_{mm}^n + \delta_m^2 I_m & k_{m*}^n \\ k_{*m}^n & k_{**} \end{bmatrix}\right) \tag{2.21}$$

where $k_{mm}^{n} = \Sigma_{m}^{n}, k_{m*}^{n} = k([x_m, x_n], x^*)$.

Likewise, the posterior predictive distribution $y^* \left| y_m^n \right.$ of y^* is the Gaussian distribution with mean m and covariance $\sum$.

$$\mu = k_{*m}^{n}(k_{mm}^{n} + \delta_m^2 I_m)^{-1}\left(y_m^n - \mu_m^n\right) + \mu_* \tag{2.22}$$

$$\Sigma = k_{**} - k_{*m}^{n}(k_{mm}^{n} + \delta_m^2 I_m)^{-1} k_{m*}^{n} \tag{2.23}$$

2.2.4 GPR-Based Fault Prediction

This section focuses on predicting the HI curve, enabling the determination of the time of fault occurrence in an unsupervised manner (when the HI value reaches zero). Utilizing normalized historical observations of HI and the corresponding operating time, the GPR-based HI predictor aims to forecast future r-step-ahead values:

$$\widetilde{h}_{l+r} = GPR(l + r) \tag{2.24}$$

where r is the prediction step size and l is the prediction starting point. Note that in general, the cases for $r = 1$ and $r > 1$ are respectively named one-step-ahead and multi-step-ahead prediction.

In our prior publication, we introduced an enhanced GPR model incorporating a linear mean function coupled with a Neural Network Squared Exponential (NNSE) kernel function. The findings indicate that this innovative model demonstrates a reduced uncertainty representation. Consequently, the mean function $\mu(x)$ and kernel function $k(x, x^*)$ are respectively selected as the linear function and NNSE covariance function, defined as follows:

$$\mu(x) = \sum_{d=1}^{p} a^d x^d + b \tag{2.25}$$

$$\begin{aligned} k(x, x^*) = {} & \delta_{NN}^2 \sin^{-1}\left(\frac{x^T x^*}{\sqrt{\left(l_{NN}^2 + x^T x\right)\left(l_{NN}^2 + x^{*T} x^*\right)}}\right) \\ & + \delta_{SE}^2 \exp\left(\frac{-(x - x^*)^2}{2 l_{SE}^2}\right) \end{aligned} \tag{2.26}$$

The recursive approach is recognized as the most straightforward and intuitive method for multi-step-ahead prediction. This technique involves repeatedly executing one-step-ahead predictions until the health indicator (HI) value reaches zero. However, it is well-known that multi-step-ahead predictions are prone to increasing uncertainty and accumulated errors as the step size r advances. Additionally, a significant amount of local fluctuations—resulting from measurement noise, environmental factors, and varying operating conditions—can be observed throughout the entire lifecycle of systems. These minor fluctuations present considerable challenges to the accurate prediction of nonlinear HI time series and fault prognosis. Directly modeling the HI time series may lead to unrealistic degradation trends, which can hinder fault prediction performance. Therefore, a prediction framework based on a moving window is proposed to enhance accuracy and robustness in fault prediction.

The moving window approach provides the most recent data to improve prognosis precision and has been utilized for continuous-time forecasting. The framework for GPR-based fault prediction is illustrated in Fig. 2.1.

For a given input set $[\widehat{h}_1, \widehat{h}_2, \ldots, \widehat{h}_{l-1}]$ and an output set $[\widehat{h}_{L+1}, \widehat{h}_{L+2}, \ldots, \widehat{h}_l]$ within the training set, the GPR-based trainer is defined using a moving window with a step size of L. Following the training process, the GPR-based HI predictor is established. As illustrated in Fig. 2.1b, the prediction value $\widetilde{h}_{l+1}$ for cycle $l+1$ is derived by inputting the known set $[\widehat{h}_{l-L}, \widehat{h}_{L-L+1}, \ldots, \widehat{h}_l]$ into the established GPR model. When the HI value at the current prediction point is obtained, the single-point update mechanism is activated, allowing for an update of the predictor.

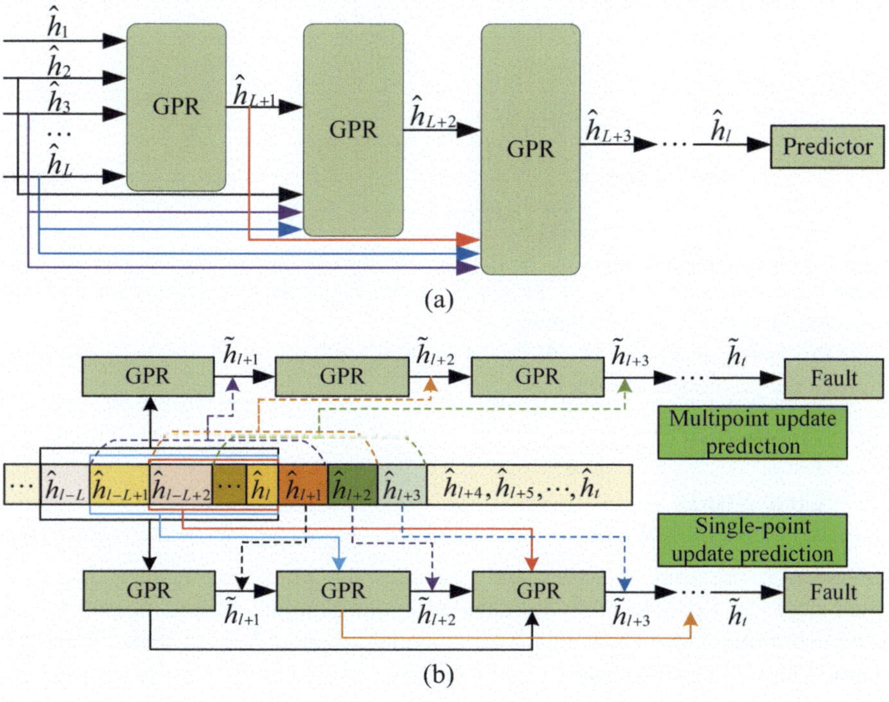

Fig. 2.1 The framework of GPR-based fault prediction. **a** Trainer. **b** Predictor

If all HI values preceding the current prediction point are available, fault prediction can be executed using the GPR model with a multipoint update strategy.For instance, when the single-point update strategy is activated at cycle $l+3$, the input set $[\widehat{h}_{l-L+2}, \ldots, \widehat{h}_l, \widetilde{h}_{l+1}, \widetilde{h}_{l+2}, \widetilde{h}_{l+3}]$ is updated to $[\widehat{h}_{l-L+2}, \ldots, \widehat{h}_l, \widetilde{h}_{l+1}, \widetilde{h}_{l+2}, \widehat{h}_{l+3}]$. Similarly, if the multipoint update mechanism is triggered at that moment, the input set is modified to $[\widehat{h}_{l-L+2}, \ldots, \widehat{h}_l, \widehat{h}_{l+1}, \widehat{h}_{l+2}, \widehat{h}_{l+3}]$. Concurrently, the GPR-based fault prediction model is updated accordingly. The cycle at which the predicted HI reaches 0 is considered the point at which the fault occurs.

To assess the prediction performance of the proposed method, we utilize the following metrics: root mean square error (RMSE), mean absolute error (MAE), mean absolute percentage error (MAPE), and score (S).

$$RMSE = \sqrt{\frac{1}{N}\sum_{i=1}^{N} e_i^2} \tag{2.27}$$

$$MAE = \frac{1}{N}\sum_{i=1}^{N} |e_i| \tag{2.28}$$

$$MAPE = \frac{1}{N}\sum_{i=1}^{N} \left|\frac{e_i}{F_{i,rea}}\right| \tag{2.29}$$

$$S = \begin{cases} \sum_{i=1}^{N} \left[exp(-\frac{e_i}{10}) - 1\right], e_i \leq 0 \\ \sum_{i=1}^{N} \left[exp(\frac{e_i}{13}) - 1\right], e_i > 0 \end{cases} \tag{2.30}$$

where the residual $e_i = F_{i,rea} - F_{i,pre}$, $F_{i,rea}$ and $F_{i,pre}$ are respectively the real and predicted fault cycle of the ith unit.

At this point, an updated GPR-based method for fault prediction is proposed to improve the prediction accuracy.

2.3 Case Study

This section primarily analyzes and discusses the prediction results. The proposed method is experimentally validated using the National Aeronautics and Space Administration (NASA) benchmark Commercial Modular Aero-Propulsion System Simulation (C-MAPSS) dataset.

2.3.1 *C-MAPSS Dataset*

The effectiveness and superiority of the applied method are validated using a case study on an aircraft turbine engine. this chapter employs the C-MAPSS dataset developed by NASA. The dataset comprises simulations of hundreds of engines operating normally but experiencing varying degrees of wear. The observation time series collected from these simulations monitor engine performance. In this chapter, we focus on the FD001 subset, which pertains to a single fault mode and operational condition. The FD001 subset includes data from 100 engines, amounting to 20,631 observations. Each engine is monitored using 58 signals, of which 21 are valid variables that record the entire degradation process from normal operation to fault state. These variables, recorded by sensors distributed across different locations and by the thermodynamic model, primarily include measurements of temperature, pressure, and speed. Table 2.1 provides detailed descriptions of these 21 variables. Given the comprehensive degradation process from health to fault state and the multivariate nature of the signals with different fault dependencies, the C-MAPSS dataset serves as a valuable resource for constructing HI to characterize the system's health state and to verify the effectiveness of the proposed method.

2.3.1.1 Variable Selection and HI Construction

The dependence between each variable and health state is quantified using the KLD. Based on a CCR threshold of 0.95, the state-dependent variables are selected by the proposed methodology. A higher threshold may result in the inclusion of irrelevant variables and noise, while a lower threshold could lead to the exclusion of relevant variables, thereby providing only partial health information. The KLD values of each variable and their CCR are illustrated in Fig. 2.2. As evident in Fig. 2.2, the KLD values of certain variables, such as variable 19, are nearly zero, indicating their minimal contribution to health information due to insignificant changes throughout the life cycle. Additionally, the CCR for variables 9 to 13 in Fig. 2.2 exceeds 0.95. Consequently, the top 13 variables are identified as the optimal state-dependent variables.

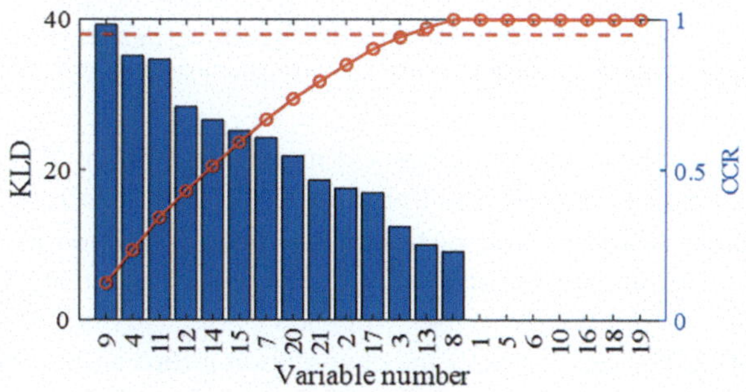

Fig. 2.2 Variables selection

Table 2.1 Detailed description of 21 variables in C-MAPSS dataset

Variable number	Description	Units
1	Total temperature at fan inlet	°R
2	Total temperature at low pressure compressor inlet	°R
3	Total temperature at high pressure compressor inlet	°R
4	Total temperature at low pressure turbine inlet	°R
5	Pressure at fan inlet	psia
6	Total pressure in bypass-duct	psia
7	Total pressure at high pressure compressor outlet	psia
8	Physical fan speed	rpm
9	Physical core speed	rpm
10	Engine pressure ratio	–
11	Static pressure at high pressure compressor outlet	psia
12	Ratio of fuel flow to Ps30	pps
13	Corrected fan speed	rpm
14	Corrected core speed	rpm
15	Bypass ratio	–
16	Burner fuel-air ratio	–
17	Bleed enthalpy	–
18	Demanded fan speed	rpm
19	Demanded corrected fan speed	rpm
20	High pressure turbine coolant bleed	lbm/s
21	Low pressure turbine coolant bleed	lbm/s

After variable selection, the weight matrix can be obtained from Eq. (2.15). Table 2.2 presents the resulting weights. For more precise HI construction, the state-dependent variables obtained are smoothed according to Eq. (2.13). The construction of HI, which is based on multivariable fusion and prediction performance, is influenced by the moving window width L. Thus, the relationship between the constructed HIs under varying window widths and the health state of engines is analyzed using equations Eqs. (2.17) to (2.19), as illustrated in Fig. 2.3. Consequently, the window width is set to 15.

According to the proposed method in section II, the HI can be constructed, as shown in Fig. 2.4. The transition of HI from 1 to 0 over time effectively represents the progression of a system or equipment from a normal operational state to a fault condition. As depicted in Fig. 2.4, the constructed HI demonstrates a clear nonlinear and monotonic decrease with each cycle. This behavior suggests that the proposed HI effectively captures the degradation information of the engine, thereby accurately

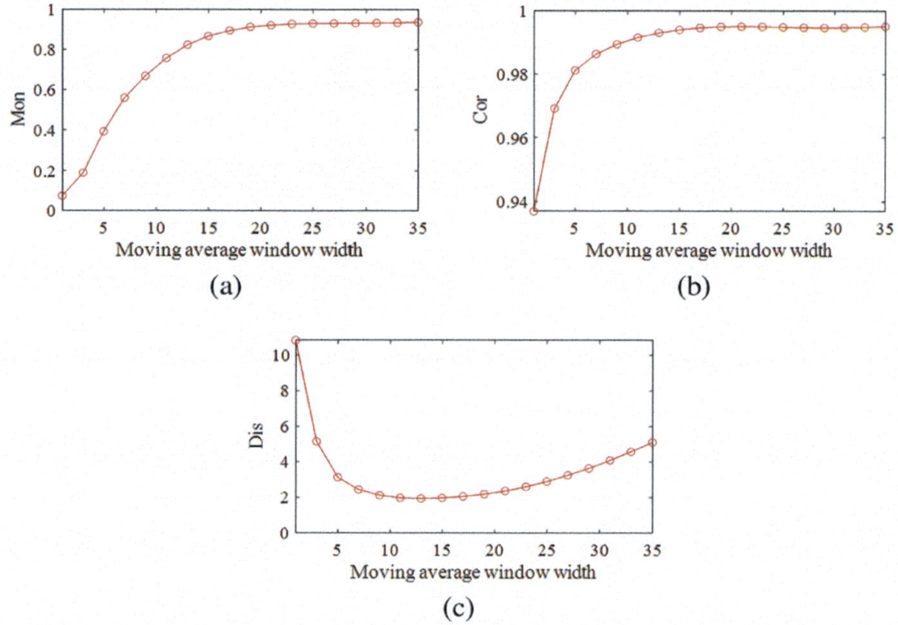

Fig. 2.3 The relationship between moving average window width and different performance criteria. **a** Mon, **b** Cor, **c** Dis

Table 2.2 The weight values of the state-relevant variables

Variable number	Weight symbol	Weight value	Variable number	Weight symbol	Weight value
9	w1	0.1263	20	w8	0.0702
4	w2	−0.1130	21	w9	0.0600
11	w3	−0.1114	2	w10	−0.0564
12	w4	0.0912	17	w11	−0.0545
14	w5	0.0856	3	w12	−0.0400
15	w6	−0.0811	13	w13	−0.0321
7	w7	0.0781			

reflecting its health state. Consequently, it is justified to consider the proposed HI as a reliable indicator of engine degradation.

To demonstrate the competitiveness of the constructed HI, a comparative analysis with several established HI construction methodologies has been conducted. These methodologies include a composite HI, a semi-parametric HI developed through a systematic data-level fusion approach, and an HI derived from a deep belief network. The comparative results, as documented in Table 2.3, indicate that the proposed HI

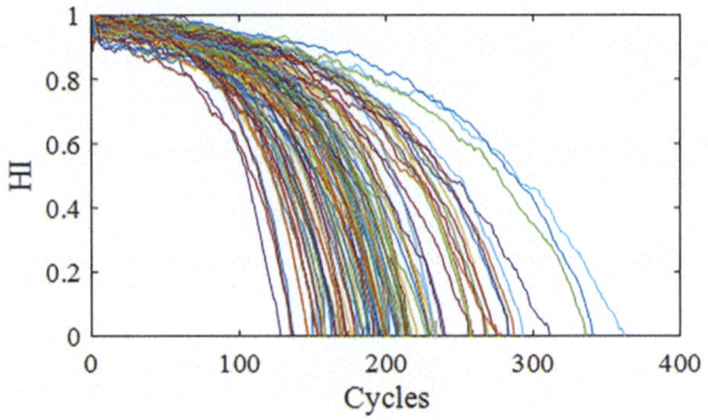

Fig. 2.4 HI construction based on the proposed method

Table 2.3 The results of HI construction by different methods

Method	Mon	Cor	Dis
Our	0.8680	0.9941	1.9797
Composite HI	0.3536	−0.9748	7.7851
Hi-semi	0.4713	−0.9879	2.3997
DBN-HI	0.7422	–	–

exhibits superior Mon, Cor, and Dis values. Consequently, the proposed HI effectively captures and reflects the latent information pertinent to the system's health state.

2.3.1.2 Fault Prediction Results and Discussion

The primary objective of the fault prediction model is to quantitatively derive the future HI curve and predict the fault occurrence time. An HI that closely approximates the actual value will more effectively reflect the degradation level and evaluate the evolutionary trend of a system or equipment. Furthermore, a fault occurrence time closer to the actual one facilitates achieving maintenance goals and improving economic benefits. To elucidate the impact of the update strategy on prediction accuracy, we compare the traditional GPR model with the GPR model that lacks an update strategy. Using engine 52 as an example, the comparison results are illustrated in Fig. 2.5, where the 95% confidence interval (CI) reflects the level of uncertainty representation. A lower 95% CI indicates reduced uncertainty and higher reliability of prediction results.

For engine 52, the traditional GPR model can only achieve short-term HI prediction with high uncertainty, failing to provide fault prediction information. In contrast, the traditional model achieves fault prediction with lower uncertainty. However, without the update mechanism, the prediction error is larger, at 26 cycles. When update strategies are incorporated, the prediction performance improves significantly for engine 52 at the prediction start point $l = 39$ (the top 65% of cycles). The errors

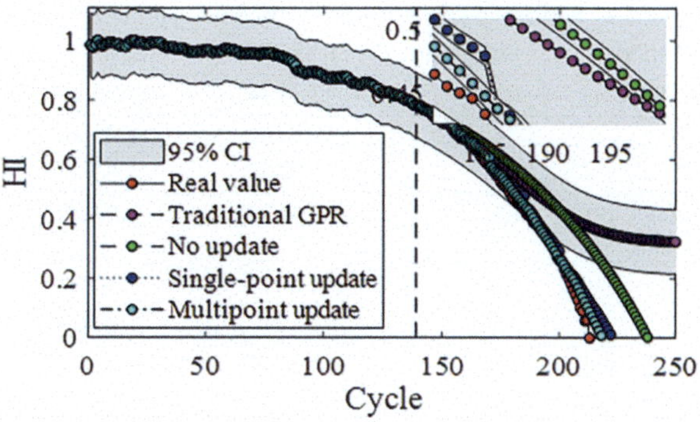

Fig. 2.5 Comparison results with and without update strategy

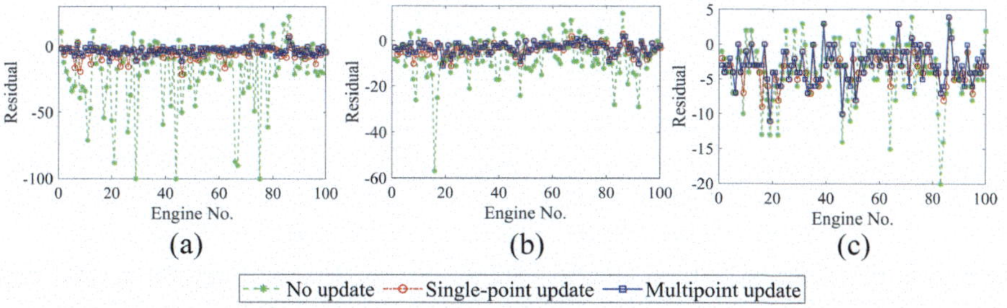

Fig. 2.6 Fault prediction results in different start points. **a** 65% cycles, **b** 75% cycles, **c** 85% cycles

decrease to 10 and 6 cycles for the single-point and multipoint update strategies, respectively. From Fig. 2.5, it is evident that HI prediction results with update strategies are closer to the actual values, allowing for quantitative evaluation of degradation levels and accurate prediction of degradation trends.

Moreover, distinct turning points are apparent in the predicted HI curves under update strategies. The HI curve predicted by the multipoint update-strategy-based GPR is closer to the actual HI curve than that obtained by the single-point update-strategy-based GPR. This improvement is due to the incorporation of more useful information for fault prediction under the update strategy, allowing the prediction parameters of the update model to better reflect the actual situation. Specifically, the multipoint update model encompasses more information about the actual situation, resulting in the most accurate fault prediction information under the multipoint update strategy.

The predicted fault time results serve as a foundation for continuous operation and maintenance downtime scheduling. To further demonstrate the efficacy of the proposed update strategy for fault prediction, all engines are evaluated using GPR both with and without the update strategy. The prediction initiation points are set at the top 65%, 75%, and 85% of the operational cycles. The prediction outcomes and performance comparisons at these different initiation points are depicted in Fig. 2.6 and summarized in Table 2.4, respectively. It is apparent that the residuals

Table 2.4 Performance comparison with and without update strategy in different start points

Method	Start point (%)	RMSE	MAE	MAPE	S
lNo update	65	33.3144	23.0300	0.1144	9104
	75	11.9796	9.0300	0.0444	494.3681
	85	5.9059	4.6200	0.0223	71.4430
lSingle-point update	65	6.5483	5.0200	0.0251	83.9816
	75	4.3543	3.6000	0.0179	47.8923
	85	4.1917	3.4900	0.0175	45.5056
lMultipoint update	65	4.2391	3.4100	0.0169	45.0229
	75	3.9077	3.1700	0.0158	41.0752
	85	3.8626	3.1600	0.0158	40.4323

gradually decrease as the training data volume increases. However, the fault prediction performance is significantly enhanced when the update mechanism is activated. Particularly for multipoint updates, the maximum residual at 85% cycles is 11. From Table 2.4, it is evident that the update mechanism enhances fault prediction performance, especially for early fault prediction. For instance, the RMSE decreases from 33.3144 to 4.2391 at 65% cycles. Furthermore, the prediction performance using the multipoint update strategy is optimal regardless of the starting point.

To demonstrate the effectiveness of the proposed method, a comparison is made among three distinct models: the BP network, the Grey model, and the RBF network, all evaluated under the same update strategy. The BP network consists of three hidden layers, utilizes Bayesian regularization as its training function, and allows for a maximum of 1000 epochs. For the Grey model, the parameters are set with a development coefficient of 0.016 and a grey action quantity of 1.337×10^{-5}. In the case of the RBF network, the target for mean squared error is 0, and the spread of the radial basis functions is specified as 10. The results of this comparative analysis are summarized in Table 2.5. As shown, all models demonstrate satisfactory performance in fault prediction overall. Among the three approaches, the BP network achieves the highest prediction accuracy within the update framework. However, it is important to note that both the BP and RBF networks do not successfully predict faults for every engine. The prediction counts for the BP and RBF networks at 65%, 75%, and 85% cycles are 72, 76, 63 and 73, 94, 81, respectively. When comparing these results with those in Table 2.4, it becomes clear that the proposed method exhibits lower error rates than the conventional time series prediction methods, highlighting the advantages of employing GPR over the alternatives.

Table 2.5 Comparison of three models on the fault prediction performance under update strategy

Method	Start point (%)	RMSE	MAE	MAPE	S
IBP	65	5.2823	4.2639	0.0204	33.9499
	75	5.2591	4.2105	0.0203	33.4011
	85	6.2640	4.3175	0.0200	44.6850
IGrey	65	9.6094	6.8600	0.0332	164.6371
	75	8.2801	5.7200	0.0282	121.5403
	85	11.8174	8.7300	0.0415	261.7290
IRBF	65	26.0771	20.7808	0.1293	647.6699
	75	15.6895	11.1383	0.0565	305.2791
	85	7.0326	5.0617	0.0237	51.9738

The performance comparisons among state-of-the-art methods are presented in Table 2.4. Detailed descriptions of these state-of-the-art methods are as follows. The MMoE-BiGRU method integrates the bidirectional gated recurrent unit (BiGRU) with the multi-gate mixture-of-experts (MMoE). A two-layer BiGRU serves as the shared bottom layer. In the MMoE structure, eight expert networks of Artificial Neural Networks (ANNs) with a four-node fully connected hidden layer are employed. The network weights are initialized using VarianceScaling, and the Adam optimizer is used with a learning rate of 0.001.

The shared bottom dual-task BiGRU (SB-BiGRU) is a conventional dual-task framework with shared BiGRU bottom layers. As a highly non-convex network structure, the bidirectional Long Short Term Memory (BD-LSTM) can track system degradation behavior by leveraging the long-term dependencies embedded in the LSTM to capture internal relationships in time series data. The batch size, learning rate, and number of epochs are 50, 0.0015, and 140, respectively, with each layer consisting of 145 neurons.

The hybrid-based method utilizes a data-decision-level hybrid fusion strategy. In this approach, a composite health index (HI) is generated through genetic programming. The results from sensors and the composite HI are integrated using belief function theory. The HI_{convex} method applies a shape-constrained neural nonlinear data fusion technique, incorporating monotonicity and convexity in the construction of the HI. The initial learning rate for this method is set to 0.01, accompanied by an exponential penalty function. The coefficients for monotonicity and convexity penalties are 2.7 and 4.2, respectively.

Table 2.6 Comparison with the existing methods reported in literature

Method	RMSE	MAE	MAPE	S
MMoE-BiGRU [8]	4.5818	–	–	54.0300
SB-BiGRU [8]	5.0042	–	–	54.4000
BD-LSTM [12]	15.4200	–	–	–
Hybrid-based [13]	24.3100	–	0.0820	–
HI_{convex} [7]	–	–	0.0450	–
Dual-task LSTM [14]	12.2900	–	–	–
LR-ED [15]	12.8062	10	0.0500	256
GP-PR [16]	30.9500	–	0.1082	–

The dual-task LSTM employs a deep LSTM architecture with peephole cell structures, specifically designed to address time-series issues. This model, consisting of two layers, is trained using the Adam optimizer, starting with a learning rate of 0.001. In the LSTM-based encoder-decoder utilizing linear regression (LR-ED), the unsupervised HI estimates the health state, bypassing the need for a knowledge-based model. Key parameters such as the number of principal components, LSTM units in the hidden layers of both the encoder and decoder, window length, maximum allowed time-lag, and similarity threshold are set to 3, 30, 20, 40, and 0.87, respectively.

The genetic programming combined with parameter reconstruction (GP-PR) method serves as an integrated prediction strategy that leverages nonlinear data-level fusion to enhance the composite HI. The tuning parameter associated with the log actual contribution ratio of the fusion result is 0.52. Most comparison methods did not yield MAE values. According to Table 2.6, the optimal RMSE value across all comparison methods is 4.5818. Given the time-varying dynamics of the system operation, the developed GPR approach with an update strategy shows superior performance compared to the other techniques.

In this chapter, the impact of the update location of newly acquired samples on the prediction performance of the proposed method is examined. During the actual prediction process, the update location is dictated by the position of the newly collected samples. Typically, these samples are gathered through equal interval sampling. Consequently, this chapter explores three distinct update intervals: long, medium, and short, with corresponding interval lengths of 40, 20, and 5 cycles, respectively. Figure 2.7 illustrates the prediction performance associated with different interval lengths under the multipoint strategy. The label C_{65}^{40} signifies that the starting point is at 65% of the cycles, with an interval length of 40 cycles. Evidently, all interval lengths achieve satisfactory prediction performance, with shorter intervals yielding superior results. Furthermore, for a given interval length, the performance metrics exhibit no significant variation with the increase in historical data. This phenomenon is attributed to the early prediction model (65% cycles) attaining comparable predictive accuracy to

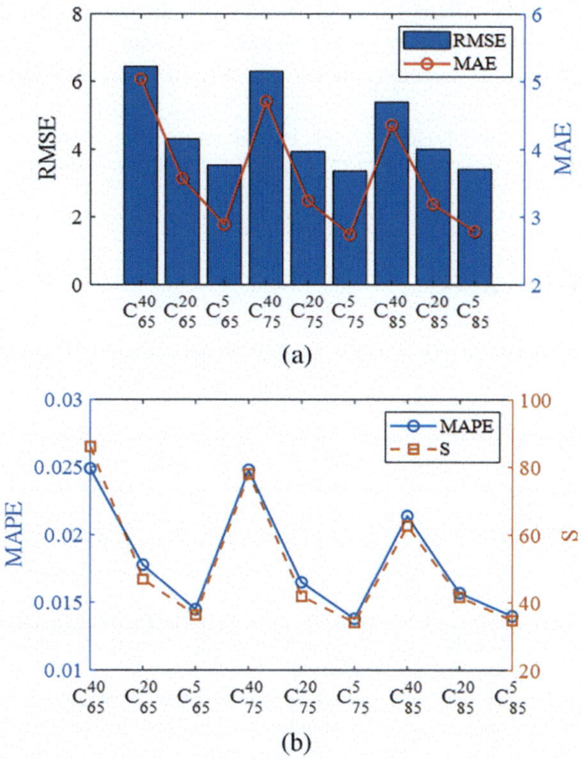

Fig. 2.7 The effect of update location of new samples on prediction performance. **a** RMSE and MAE, **b** MAPE and S

the late model (85% cycles) after several update steps. This observation substantiates the robustness of the proposed update-strategy-based GPR model.

2.4 Conclusion

In this chapter, we present a GPR-based prognostic method with an update strategy for the fault prediction of aeroengines. Initially, optimal relevant variables are identified through KLD-based variable selection. Given the labor-intensive nature of manually designed labels, we introduce an unsupervised HI construction approach utilizing multivariable fusion based on KLD weight setting, which effectively characterizes the condition of a system or unit. Subsequently, the distributions for new samples are inferred to ensure accurate and robust prediction results. The proposed framework concurrently achieves HI and fault time prediction, thereby eliminating the need for manually balancing weights between the dual tasks. Experimental results demonstrate that the proposed method exhibits superior prediction performance compared to existing methods, providing a valuable reference for subsequent maintenance decisions. Notably, the GPR model with a multipoint update strategy outperforms the single-point-based model. Furthermore, the robustness of the proposed multipoint update strategy-based GPR model is validated through interval length analysis.

References

1. Ragab M, Chen Z, Wu M, Foo CS, Kwoh CK, Yan R, Li X (2020) Contrastive adversarial domain adaptation for machine remaining useful life prediction. IEEE Trans Indus Inform 17(8):5239–5249
2. Miao H, Li B, Sun C, Liu J (2019) Joint learning of degradation assessment and rul prediction for aeroengines via dual-task deep lstm networks. IEEE Trans Indus Inform 15(9):5023–5032
3. Peng X, Wei G, Song K, Chen Y (2021) High-accuracy health prediction of sensor systems using improved relevant vector-machine ensemble regression. Knowl Based Syst 212:106555
4. Chen S, Wang M, Huang D, Wen P, Wang S, Zhao S (2020) Remaining useful life prediction for complex systems with multiple indicators based on particle filter and parameter correlation. IEEE Access 8:215145–215156
5. Soleimani M, Campean F, Neagu D (2021) Integration of hidden markov modelling and bayesian network for fault detection and prediction of complex engineered systems. Reliab Eng Syst Safety 215:107808
6. Duan C, Li Y, Huayan P, Luo J (2022) Multi-attribute bayesian fault prediction for hidden-state systems under condition monitoring. Appl Mathem Model 103:388–408
7. Li Z, Jianguo W, Yue X (2020) A shape-constrained neural data fusion network for health index construction and residual life prediction. IEEE Trans Neural Netwo Learn Syst 32(11):5022–5033
8. Zhang Y, Xin Y, Liu Z, Chi M, Ma G (2022) Health status assessment and remaining useful life prediction of aero-engine based on bigru and mmoe. Reliab Eng Syst Safety 220:108263
9. Liu K, Huang S (2014) Integration of data fusion methodology and degradation modeling process to improve prognostics. IEEE Trans Autom Sci Eng 13(1):344–354
10. Deng L, Shen W, Wang H, Wang S (2021) A rest-time-based prognostic model for remaining useful life prediction of lithium-ion battery. Neural Comput Appl 33:2035–2046
11. Lin C-P, Cabrera J, Yang F, Ling M-H, Tsui K-L, Bae S-J (2020) Battery state of health modeling and remaining useful life prediction through time series model. Appl Energy 275:115338
12. Zhang J, Wang P, Yan R, Gao RX (2018) Long short-term memory for machine remaining life prediction. J Manuf Syst 48:78–86
13. Wen P, Li Y, Chen S, Zhao S (2021) Remaining useful life prediction of iiot-enabled complex industrial systems with hybrid fusion of multiple information sources. IEEE Internet Things J 8(11):9045–9058
14. Miao H, Li B, Sun C, Liu J (2019) Joint learning of degradation assessment and rul prediction for aeroengines via dual-task deep lstm networks. IEEE Trans Indus Inform 15(9):5023–5032

15. Malhotra P, Vishnu TV, Ramakrishnan A, Anand G, Vig L, Agarwal P, Shroff G (2016) Multi-sensor prognostics using an unsupervised health index based on lstm encoder-decoder. arXiv preprint arXiv:1608.06154
16. Wen P, Zhao S, Chen S, Li Y (2021) A generalized remaining useful life prediction method for complex systems based on composite health indicator. Reliab Eng Syst Safety 205:107241

Chapter 3
Relevant Variable Selection and SVDD for Processes Fault Detection with Incremental Redundant Data

This chapter addresses the issue of sample value imbalance in process monitoring by developing a fault detection approach that integrates variable selection and support vector data description (SVDD) for enhanced efficiency. Initially, KLD is employed as the variable selection algorithm to identify the most informative variables related to specific faults. These selected variables are then segmented into blocks to prevent fault information from being obscured within a single monitoring space, ensuring that relevant variables and critical information are concentrated within the same block. Subsequently, kernel principal component analysis is conducted within each block to manage the challenges posed by high-dimensional and nonlinear variables. The monitoring outcomes are derived using the proposed SVDD, which incorporates a restructured radius index that exhibits heightened sensitivity to faults. Experimental results on the Tennessee Eastman process demonstrate the efficacy of this method, highlighting its superior mean fault detection rate compared to existing approaches.

3.1 Motivation

SVDD has emerged as a powerful tool for data-driven fault detection, gaining widespread acceptance in various fields, including rolling element bearings [1], aircraft engines [2], and chillers [3]. However, existing SVDD methods often overlook the irregularities of the hypersphere, using the mean distance from all support vectors to the center as the detection boundary radius, which results in suboptimal detection performance. In contrast, conventional MSPM techniques such as principal component analysis (PCA), canonical correlation analysis, Fisher discriminant analysis, and partial least squares, while effective for extracting latent features, face limitations in process monitoring due to challenges in selecting relevant information for large-scale industrial processes, reducing monitoring redundancy, and accurately describing faults.

H. Yin et al., *Data-Driven Fault Diagnosis for Complex Industrial Processes*,
Engineering Applications of Computational Methods 22,
https://doi.org/10.1007/978-981-96-3153-7_3

In large-scale processes, the number of process variables is on the rise, and their interconnections are complex [4]. When faults occur, not all variables are relevant [5]. In particular, many variables may have weak or irrelevant connections to the fault in question, while only a few are truly related [6], leading to an imbalance in sample values. Monitoring all variables within a single information space may result in poor performance due to overlooked fault information [7]. To improve monitoring effectiveness, multiblock monitoring algorithms have been developed. For example, Zhu et al. [8] proposed a multiblock decoupled convolutional neural network aimed at key variable selection for large-scale quality-related process monitoring. Another study introduced a multiblock kernel PCA (KPCA) approach that accounts for both high-order data statistics and nonlinearity, proving effective for process monitoring [9]. Despite the strengths of these methods, monitoring performance may still be suboptimal if fault-related information is not fully leveraged [7], and unrelated variables may unnecessarily increase computational costs. Therefore, it is essential to identify and select fault-relevant variables that capture the most significant effects of faults, enhancing understanding and interpretation of the fault process.

Notably, it is uncertain whether variable selection alone suffices to derive a low-dimensional optimal subset from the original data in large-scale, high-dimensional industrial settings. The complexity of industrial processes and the interdependence of variables complicate the identification of those relevant to faults. Fault occurrences often lead to deviations in multiple variables from their normal states, meaning that subsets selected solely through variable selection methods may still be high-dimensional and may include variables unrelated to the fault. Moreover, the correlations among different variables in the selected subset can vary, potentially obscuring crucial fault-related information during monitoring.

To tackle these challenges, this chapter proposes an enhanced SVDD approach that incorporates a novel monitoring index following variable selection, block segmentation, and feature extraction. The variable selection algorithm based on KLD is employed to prioritize the most informative variables and eliminate irrelevant ones. The selected variables are then segmented into blocks based on KLD results and empirical knowledge, ensuring that relevant variables and critical information are grouped together to avoid being overlooked in single-space monitoring. KPCA is applied to each block to manage nonlinearity, computational costs, and high dimensionality. The monitoring processes are then restructured using a moving window approach. An improved SVDD method, with a restructured radius index, is developed to produce final monitoring outcomes in complex industrial environments.

The main contributions of this chapter are as follows:

- A KLD-based block segmentation scheme is proposed to address the issue of sample value imbalance in fault detection. Variables exhibiting significant fault-related KLD values are identified and grouped into blocks, thereby reducing dimensionality while preserving crucial fault information.
- Feature extraction within each block employs KPCA to capture nonlinear features inherent in process data. The block-based KPCA approach facilitates simultaneous

feature extraction across multiple subspaces, further reducing dimensionality and computational cost associated with nonlinear feature extraction.

- A fault detection method leveraging SVDD is introduced to enhance monitoring performance. The proposed method, utilizing a restructured radius index, amalgamates monitoring statistics from different blocks, thereby minimizing misclassification rates and delivering competitive results owing to its heightened fault sensitivity.

3.2 Methodology

3.2.1 Variable Selection, Block Division and Feature Extraction

To mitigate the impact of outliers and the varying feature scales across different dimensions in the raw data, normalization is essential. In this context, the z-score normalization method is employed, as outlined below:

$$x_{nor} = \frac{x - \mu(x)}{\delta(x)} \tag{3.1}$$

where $\mu(x)$ and $\mu(x)$ are respectively the mean and standard deviation of the raw data x.

Variable selection aims to identify the most informative fault-related variables. The KLD can quantify the divergence between the distributions of a variable under normal and abnormal conditions. Variables exhibiting high KLD values are considered more strongly associated with faults. Consequently, KLD serves as a metric to assess the relationship between normal and fault states for a given variable. However, variables with low KLD values may inadequately represent the dependence between variables and faults, as noise and other disturbances can also result in low KLD values. To address this limitation, the cumulative contribution rate (CCR) is introduced as follows:

$$CCR_n = \frac{\sum\limits_{i=1}^{n} KLD_i}{\sum\limits_{i=1}^{m} KLD_i} \times 100\% \tag{3.2}$$

where m represents the total number of variables. For a specified CCR limit, the number of relevant variables, denoted as n, can be established. It is important to note that the KLD results should be arranged in descending order prior to determining the variable count.

Fault-related variables are selected based on KLD in conjunction with the specified CCR limit. However, the dependency of each variable within the subset on the identified fault may differ, leading to a potentially high-dimensional variable subset. This is particularly true for incipient faults and those involving highly interrelated variables. As a result, the subset might still encompass unrelated variables, which is evidenced by substantial differences in KLD values between the most and least significant variables. Monitoring these variables in the same space may dilute the influence of those that are strongly associated with the faults. To tackle these challenges, this chapter proposes the use of block segmentation and KPCA techniques.

Since the majority of variables identified by KLD are fault-related, they can be viewed as having strong or relatively strong correlations with the fault. The simplicity of the model is also a consideration. Consequently, the variable subset is divided into two distinct blocks. The number of variables in each block is determined using KLD results alongside empirical insights. In processes characterized by weakly coupled variables, the count of variables in each block can be easily defined based on KLD results, as there are significant differences in KLD values among the variables. Conversely, in processes that change gradually or include a greater number of strongly coupled variables, the first block (which is strongly correlated with the fault) may contain slightly more variables than the second block (which has a lesser correlation). Once the variable counts for each block are established, the variable subset $F = [F_1, F_2, \ldots, F_n]$ can be partitioned into $[F_1, F_2, \ldots, F_b]$ and $\left[F_{b+1}, F_{b+2}, \ldots, F_n\right]$. Subsequently, the variables in each block undergo dimensionality reduction via KPCA, allowing the principal components to be extracted based on a specified contribution rate. The T^2 and SPE for each block can be calculated using the following expressions:

$$T^2 = [y_1, y_2, \cdots, y_k]\,\Lambda^{-1}[y_1, y_2, \cdots, y_k]^T \tag{3.3}$$

$$SPE = \sum_{i=1}^{m} y_i^2 - \sum_{i=1}^{k} y_i^2 \tag{3.4}$$

where y_k is the kth principal component after dimension reduction. Λ is the diagonal matrix with the eigenvalues of the covariance matrix $\widetilde{\Sigma}$ in the feature space.

3.2.2 Fault Detection Based on Improved SVDD

SVDD is introduced to detect the fault. In every block, KPCA is applied and the corresponding T^2 and SPE can be yielded. The obtained T^2 and SPE in every block are combined as $Y = [SPE_1, T_1^2, SPE_2, T_2^2]$ to serve as the training matrix of SVDD. It is pertinent to note that the incorporation of a limited number of fault samples into the normal sample set serves to mitigate the misclassification rate during the training phase. Figure 3.1 elucidates the training process. The red markers represent normal samples, while the green markers denote support vectors. When the training set is comprised exclusively of single-class samples, the decision boundary (represented by a solid line) is prone to overfitting, which subsequently leads to a heightened misclassification rate during testing and consequently diminishes the accuracy of process monitoring. Conversely, the inclusion of a few fault samples (illustrated as blue markers) within the training set ameliorates this issue, as depicted in Fig. 3.1. This improvement occurs because the inclusion of fault samples facilitates the clustering of as many normal samples as possible within the decision boundary (depicted by a dashed line), while the fault samples remain outside this boundary. Essentially, the training process tolerates the misclassification of a minor subset of samples to achieve an optimal decision boundary. Therefore, by integrating a small number of fault samples into the training set, an enhanced SVDD model is proposed.

Assuming that the labels of normal and abnormal samples in the training set are $z_k = +1$ and $z_k = -1$, respectively, the dual problem is

$$\begin{aligned} &\min_{\lambda_i} \sum_{k=1}^{m} \sum_{i=1}^{m} z_k z_i \lambda_k \lambda_i k_{ki} - \sum_{k=1}^{m} z_k \lambda_k k_{kk} \\ &s.t. \quad \sum_{k=1}^{m} z_k \lambda_k = 1 \\ &\qquad 0 \leq \lambda_k \leq C_1, z_k = +1 \\ &\qquad 0 \leq \lambda_k \leq C_2, z_k = -1 \end{aligned} \tag{3.5}$$

The center a and radius r of the hypersphere are denoted as follows:

$$a = \sum_{k=1}^{m} z_k \lambda_k \varphi(x_k) \tag{3.6}$$

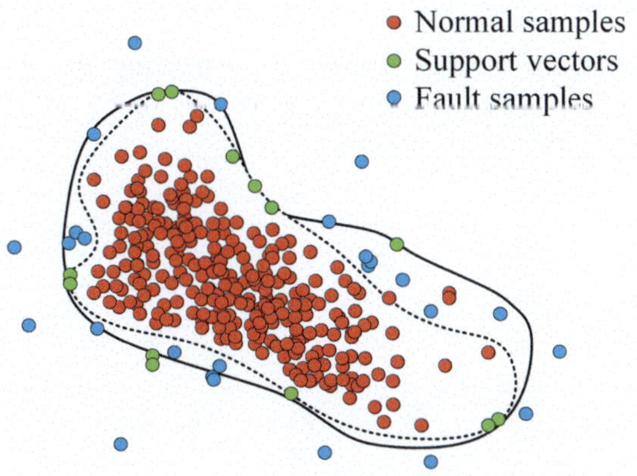

Fig. 3.1 Illustration of the training process for improved SVDD model

$$r = \sqrt{\sum_{k=1}^{m}\sum_{i=1}^{m} z_k z_i \lambda_k \lambda_i k_{ki} + k_{vv} - 2\sum_{k=1}^{m} z_k \lambda_k k_{vk}} \tag{3.7}$$

To mitigate the impact of measurement noise on monitoring, a moving window mechanism is introduced to enhance the fault detection rate (FDR) and decrease the false alarm rate (FAR). By integrating SVDD with the moving window mechanism, a novel statistical criterion is developed. This combination enhances the distinction between normal and abnormal samples, thereby increasing the sensitivity of the statistical criterion to the targeted faults. For the window length l, the samples in the window at time t can be denoted as follows:

$$Y_t^l = [Y_{t-l+1}, Y_{t-l+2}, \cdots, Y_t]^T \tag{3.8}$$

where $Y_t = [SPE_{1t}, T_{1t}^2, SPE_{2t}, T_{2t}^2]$. To reduce the noise, the training samples are restructured. Similar to the root mean square, the restructured samples can be defined by Eq. (3.9).

$$\widetilde{Y}_t^l = \sqrt{\sum_{i=1}^{l} \frac{i}{\sum_{i=1}^{l} i} Y_{t-l+i}^2} \tag{3.9}$$

According to the proposed SVDD model, the corresponding threshold $\widetilde{r}_{thre}$ can be attained based on the training set.

$$\widetilde{r}_{thre} = \max\left(\sqrt{\sum_{i=1}^{l} \frac{i}{\sum_{i=1}^{l} i} r_{tra}^2}\right) \tag{3.10}$$

where $r_{tra}^2 = \sum_{k=1}^{m}\sum_{i=1}^{m} z_k z_i \lambda_k \lambda_i k_{ki} + k_{tratra} - 2\sum_{k=1}^{m} z_k \lambda_k k_{trak}$. The sub-script tra denotes the normal samples in the training set.

Therefore, for the new sample x_{new} in the testing set, the restructured radius $\widetilde{r}_{new}$ can be defined as follows:

$$\widetilde{r}_{new} = \sqrt{\sum_{i=1}^{l} \frac{i}{\sum_{i=1}^{l} i} r_{new}^2} \tag{3.11}$$

where $r_{new}^2 = \sum_{k=1}^{m}\sum_{i=1}^{m} z_k z_i \lambda_k \lambda_i k_{ki} + k_{newnew} - 2\sum_{k=1}^{m} z_k \lambda_k k_{newk}$. As a result, if $\widetilde{r}_{thre} < \widetilde{r}_{new}$, it means that a fault occurs.

Figure 3.2 illustrates the comprehensive framework of the fault detection method employed for process monitoring. Figure 3.3 delineates the detailed procedural steps of the proposed monitoring method. Initially, the most relevant variables are identified using KLD, and the threshold is determined through an enhanced SVDD approach. During the monitoring phase, only the preselected variables are input into the constructed model, thereby enabling the calculation of the monitoring statistic. To evaluate the monitoring performance, the FDR and FAR are introduced, and they are expressed as follows:

$$FDR = 1 - \frac{\text{Fault data identified as normal}}{\text{Fault data}} \tag{3.12}$$

$$FAR = \frac{\text{Normal data identified as abnormal}}{\text{Normal data}} \tag{3.13}$$

3.3 Case Study

This section provides an analysis and discussion of the fault monitoring results. The TE benchmark process, incorporating various fault types, is utilized to validate the reliability and accuracy of the proposed method. Additionally, the impact of different window lengths on monitoring performance is examined.

3.3.1 Tennessee Eastman Process

The TEP serves as a benchmark for process monitoring and fault diagnosis. The training set comprises 500 normal operation samples and 480 abnormal operation samples due to various faults. The testing set contains 960 samples, with the first 160 representing normal operation and the remainder depicting abnormal conditions for each fault scenario.

By applying the CCR limit, it is possible to identify fault-related variables. It is important to note that setting a higher limit may lead to the inclusion of irrelevant variables, while a lower limit could result in missing relevant ones. Therefore, the CCR limit is set at 90 percentage. Utilizing KLD with this limit allows for the extraction of a subset from the original variables, as outlined in Table 3.1. Remarkably, all identified faults demonstrate a reduction in dimensionality. Specifically, for faults F4, F17, and F21, the dimensionality decreases from 51 to 1. However, incipient faults, such as F3, maintain a relatively high dimensionality due to strongly coupled variables and minor fluctuations.

To visualize the relationships between the selected variables and the corresponding faults, these variables are plotted in 2D or 3D space. Figure 3.4 illustrates the

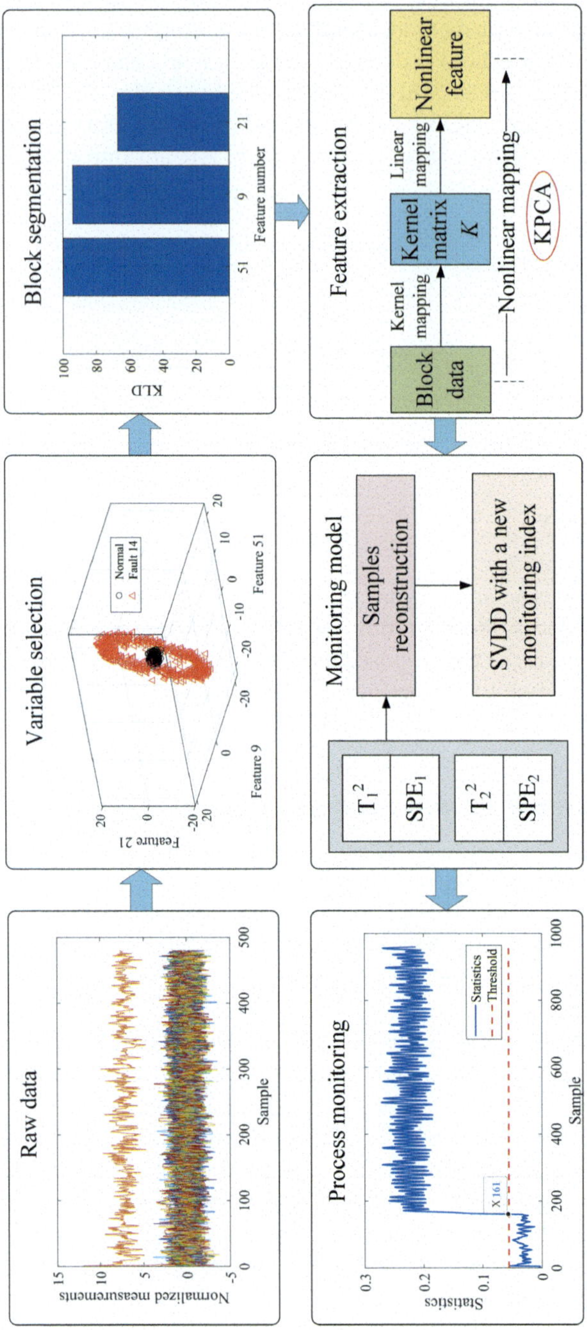

Fig. 3.2 The framework of the proposed model for process monitoring

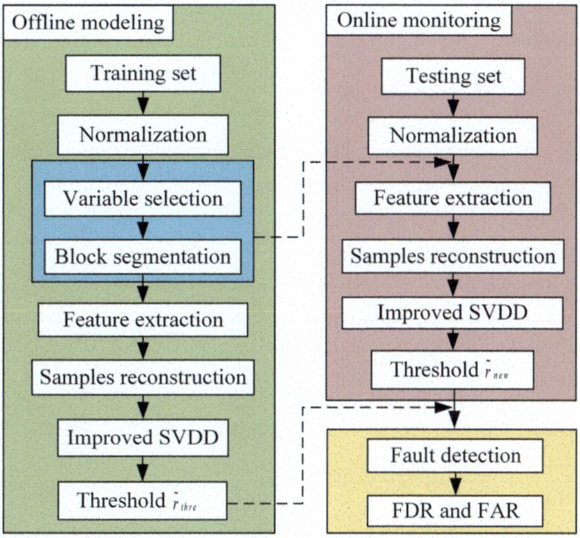

Fig. 3.3 The step-by-step procedure of process monitoring

distinction between normal and fault samples in the feature space for faults F4, F6, F11, F14, F17, and F21 from the training set. It is clear that normal and fault samples are well-separated for faults F4, F6, and F14. For faults F11, F17, and F21, most samples are also adequately distinct, suggesting the variable selection method is effective. However, it is evident that not all subsets of variables are optimal. For example, further dimensionality reduction in the feature space for fault F6 can be achieved by projecting along the direction of feature 16. In contrast, for certain faults like F3, F8, and F12, the variable subsets identified through KLD remain high-dimensional. This indicates that variable selection alone is inadequate for deriving an optimal low-dimensional subset of the original data in large-scale, high-dimensional industrial contexts. Consequently, block segmentation and KPCA are implemented to further address these challenges.

After selecting the variables, the subsets can be partitioned into two distinct blocks, with the number of variables in each block being determined by the Kullback-Leibler Divergence (KLD) results and empirical knowledge. Figure 3.5 illustrates the KLD outcomes for various fault types, namely step (F1), random (F7), slow drift (F13), sticking (F14), and unknown faults (F16). Additionally, fault F9 represents a random incipient fault. Based on Fig. 3.5, the variable numbers in the first block are determined from the KLD results. For example, the variable counts b in the first block are 2, 1, 5, 2, and 8 for faults F1, F7, F13, F14, and F16, respectively. For fault F9, however, the higher coupling among variables (indicated by small and nearly uniform KLD values) suggests that the performance may not be optimal when relying solely on KLD results. Therefore, empirical knowledge is also incorporated. Given the small differences among KLD results for incipient faults, the variable number is adjusted based on the total variable count. Specifically, for fault F9, the number of

Table 3.1 Variable subsets using KLD based on 90% CCR limit for 21 faults

Fault no.	No. of variable	Rank variable subset (90% CCR limit)
F1	14	1, 44, 50, 18, 19, 4, 16, 7, 13, 20, 31, 38, 46, 29
F2	7	10, 47, 34, 28, 19, 50, 18
F3	27	20, 21, 40, 13, 7, 3, 19, 37, 50, 39, 27, 38, 46, 16, 35, 18, 26, 42, 34, 24, 36, 43, 51, 41, 11, 14, 31
F4	1	51
F5	19	50, 19, 18, 7, 13, 16, 38, 20, 46, 11, 31, 25, 33, 52, 29, 1, 44, 4, 35
F6	3	1, 51, 16
F7	19	45, 50, 7, 20, 13, 16, 19, 38, 18, 46, 11, 31, 33, 29, 4, 27, 1, 44, 25
F8	21	16, 20, 7, 13, 31, 46, 29, 50, 19, 18, 25, 38, 11, 1, 44, 23, 4, 33, 47, 10, 35
F9	18	18, 50, 19, 16, 7, 13, 20, 29, 3, 38, 41, 31, 35, 11, 25, 46, 34, 28
F10	9	19, 50, 18, 38, 16, 20, 7, 13, 31
F11	2	51, 9
F12	21	7, 13, 16, 20, 11, 46, 50, 18, 38, 19, 22, 31, 21, 33, 4, 35, 27, 29, 25, 3, 1
F13	17	16, 50, 19, 7, 13, 18, 11, 38, 35, 46, 20, 33, 22, 21, 3, 27, 36
F14	3	51, 9, 21
F15	24	19, 50, 18, 40, 16, 34, 7, 13, 27, 39, 29, 31, 3, 38, 37, 11, 28, 12, 48, 35, 22, 9, 46, 20
F16	17	50, 19, 18, 13, 7, 20, 16, 46, 31, 11, 4, 3, 38, 40, 27, 33, 29
F17	1	21
F18	18	43, 7, 13, 16, 46, 51, 42, 20, 11, 45, 4, 38, 47, 33, 29, 19, 50, 27
F19	14	5, 46, 51, 9, 21, 3, 38, 16, 27, 13, 19, 40, 18, 7
F20	5	46, 20, 13, 11, 7
F21	1	45

variables in the first block is slightly greater than that in the second block, resulting in a variable count b of 10 for the first block.

After block segmentation, the dimensionality of the variables within each block is reduced using KPCA, considering a contribution rate of 0.9 and a window length of 6.

To demonstrate the superiority of the proposed method in monitoring performance, comparisons are made with ICA, KPCA, MI-MBPCA, and SVDD. The monitoring results are presented in Table 3.2, where the highest FDR are highlighted in bold. As evidenced, the proposed method, similar to other methods, accurately detects faults F4, F6, F7, F12, and F14. For faults F1, F2, F8, and F13, the FDRs are nearly equivalent to the best results of the other four methods, with all differences being 0.01. Moreover, the proposed method achieves the highest FDRs for 14 out of all faults, notably achieving FDRs of 0.98 and 0.99 for faults F11 and F21, respectively. Additionally, compared to the optimal performance of the other four models, the proposed method's FDRs increase by 22%, 12%, and 4% for incipient faults F3,

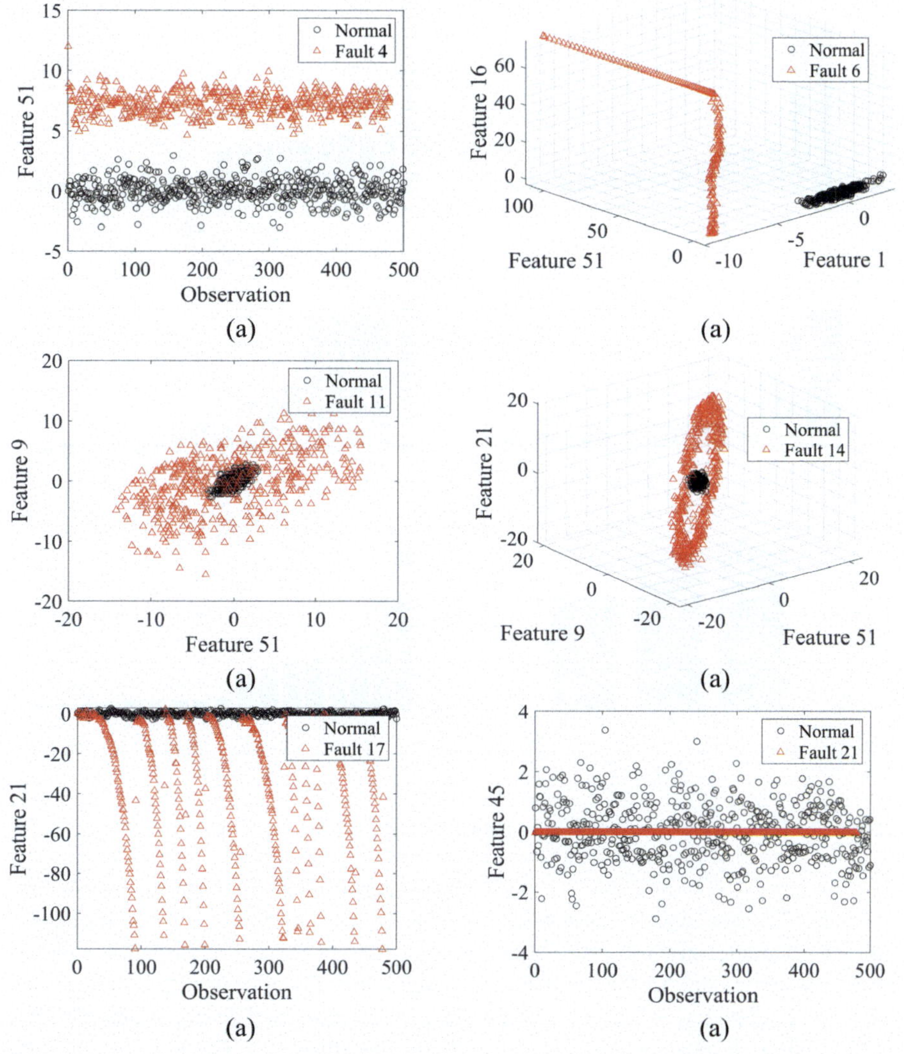

Fig. 3.4 Variable selection for different faults. **a** F4, **b** F6, **c** F11, **d** F14, **e** F17, **f** F21

F9, and F15, respectively, indicating a significant improvement in monitoring performance for incipient faults. As shown in Table 3.2, the mean FDR of the proposed method is higher, standing at 0.83. Furthermore, Table 3.2 presents the FAR of the proposed method. It is evident that most FARs are relatively small, except for faults F9 and F16, likely due to higher variable coupling, making it more challenging to identify the fault-relevant variable subset. Overall, the proposed method outperforms the others.

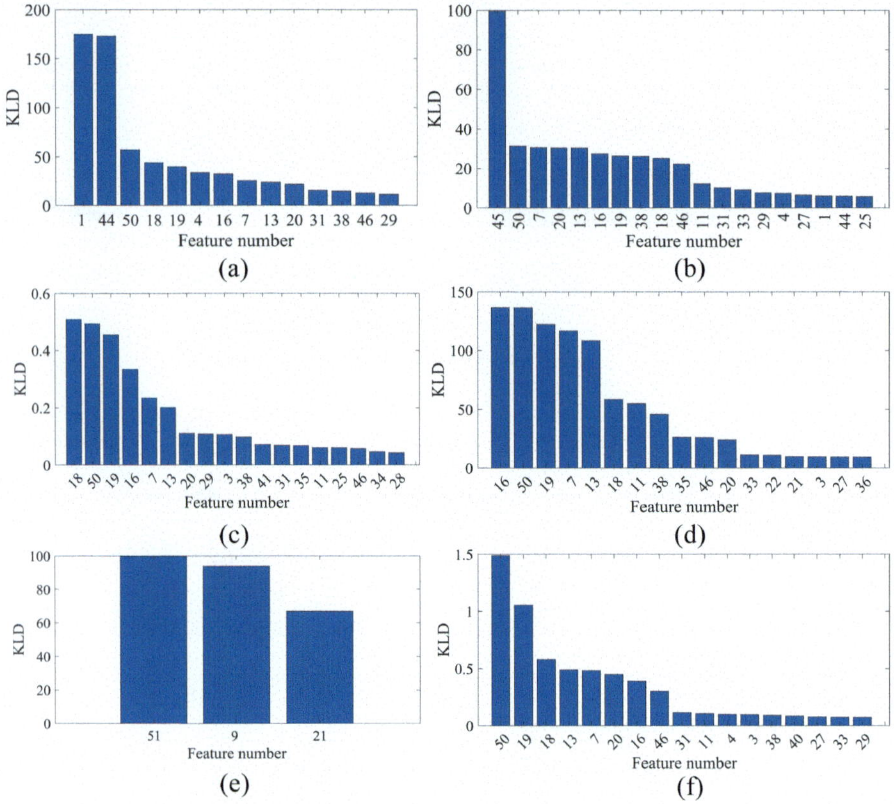

Fig. 3.5 KLD result for different types of faults: **a** F1, **b** F7, **c** F9, **d** F13, **e** F14, **f** F16

Figure 3.6 presents the monitoring results for various fault types. Specifically, fault F1, characterized by a step change in the A/C feed ratio in stream 4, induces large oscillations due to the control loop's influence. The monitoring outcomes for this fault are depicted in Fig. 3.6a, where the proposed method demonstrates a 0 FAR for fault F1, as all normal samples remain below the threshold. The fault is identified at the 171st sample. Comparable findings are observed in Fig. 3.6b, d, e, with faults detected at the 161st, 205th, and 161st samples for random fault F7, slow drift fault F13, and sticking fault F14, respectively.

What's more, the monitoring results of the incipient fault and unknown fault are also plotted in Fig. 3.6c, f. It should be highlighted that the FARs of the two faults are the highest in all 21 faults according to Table 3.3. From Fig. 3.6c, f, the monitoring results are not satisfied with those of the other faults. This may be because there are more high coupling variables, which causes the reconstructed index to be not as sensitive to these faults as other faults.

Figures 3.7 and 3.8 illustrate the monitoring results of KPCA and SVDD for various fault types, respectively. Both methods achieve a 100% FDR for faults F1,

Table 3.2 DR of different methods for 21 faults in TEP

Fault no.	ICA		MI-MBPCA		MI-MBPCA	SVDD	Our
	I^2	SPE	T^2	SPE	DR	d/r	$\widetilde{r}_{new}$
F1	1.00	1.00	1.00	1.00	1.00	1.00	0.99
F2	0.99	0.99	0.99	0.99	0.99	0.99	0.98
F3	0.03	0.08	0.08	0.08	0.09	0.20	0.42
F4	0.78	0.89	1.00	0.37	1.00	0.93	1.00
F5	1.00	1.00	0.29	1.00	1.00	0.38	0.57
F6	1.00	1.00	1.00	1.00	1.00	1.00	1.00
F7	1.00	1.00	1.00	1.00	1.00	1.00	1.00
F8	0.97	0.99	0.98	0.98	0.99	0.99	0.98
F9	0.01	0.03	0.07	0.05	0.11	0.17	0.29
F10	0.75	0.76	0.55	0.87	0.85	0.65	0.73
F11	0.70	0.55	0.79	0.52	0.78	0.74	0.98
F12	1.00	1.00	0.99	1.00	1.00	1.00	1.00
F13	0.95	0.96	0.96	0.96	0.96	0.96	0.95
F14	1.00	1.00	1.00	1.00	1.00	1.00	1.00
F15	0.04	0.15	0.13	0.14	0.19	0.24	0.28
F16	0.80	0.75	0.37	0.90	0.91	0.59	0.61
F17	0.88	0.90	0.96	0.91	0.96	0.92	0.98
F18	0.91	0.90	0.91	0.89	0.91	0.91	0.91
F19	0.65	0.37	0.19	0.81	0.74	0.19	0.97
F20	0.78	0.74	0.68	0.72	0.77	0.67	0.78
F21	0.43	0.49	0.55	0.44	0.52	0.47	0.99
Mean	0.75	0.74	0.69	0.74	0.80	0.71	0.83

Table 3.3 FAR of the proposed method for 21 faults in TEP

Fault no.	F1	F2	F3	F4	F5	F6	F7
FAR	0	0	0.20	0.03	0.03	0	0.02
Fault no.	F8	F9	F10	F11	F12	F13	F14
FAR	0	0.45	0.01	0.01	0.18	0	0
Fault no.	F15	F16	F17	F18	F19	F20	F21
FAR	0.05	0.36	0.03	0	0	0	0.01

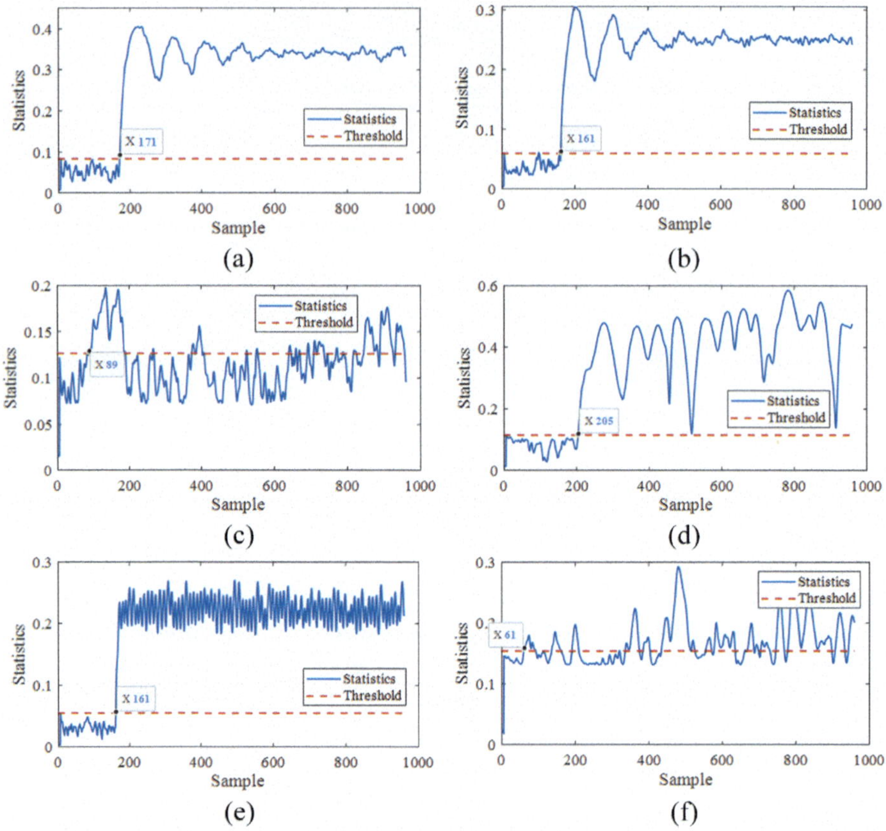

Fig. 3.6 Monitoring results of the proposed method for different types of faults: **a** F1, **b** F7, **c** F9, **d** F13, **e** F14, **f** F16

F7, and F14. However, the fault occurrence time is detected earlier by these methods compared to the proposed method. Specifically, for fault F1, it is detected at approximately the 51st and 53rd samples for KPCA and SVDD, respectively. This indicates a potential misclassification using these methods, likely due to the inability to effectively distinguish between disturbance and fault, resulting in a lower threshold. For fault F9, neither KPCA nor SVDD can detect the fault occurrence time due to the continuous fluctuation of the statistical index. In contrast, the proposed method detects the fault between approximately the 160th to 200th samples and after the 850th sample. Furthermore, as seen in Figs. 3.6f, 3.7f, and 3.8f, the detection time for fault F16 is nearly identical across the methods, occurring around the 60th sample. Overall, the proposed method demonstrates superior fault detection performance.

To illustrate the computational complexity of the proposed method, the fault detection time is presented in Table 3.4. The processing time was obtained using MATLAB 9.10, with all computations conducted on a PC equipped with an Intel Core i5-7500

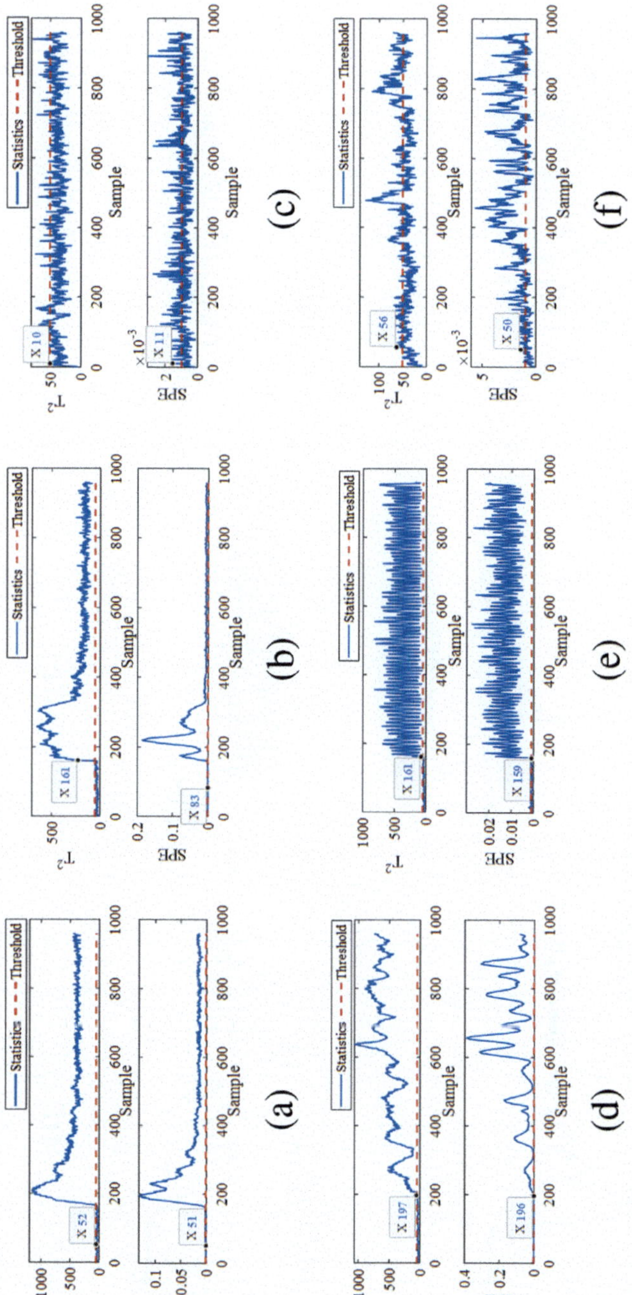

Fig. 3.7 Monitoring results of KPCA for different types of faults: **a** F1, **b** F7, **c** F9, **d** F13, **e** F14, **f** F16

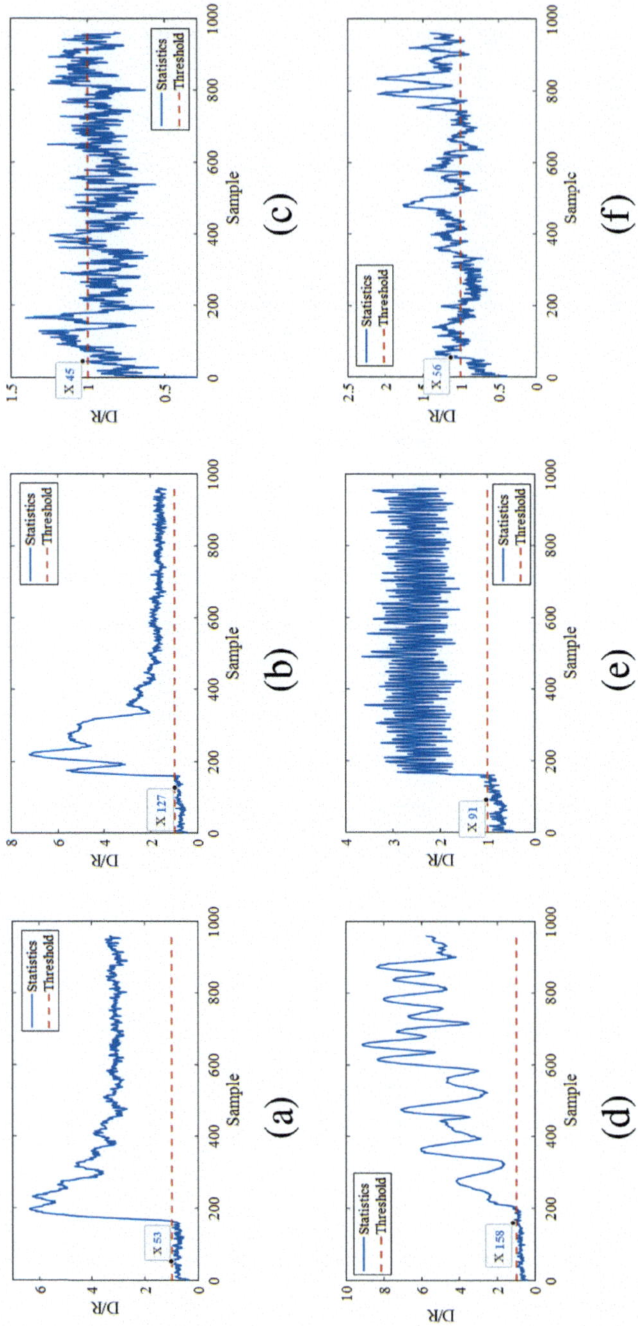

Fig. 3.8 Monitoring results of SVDD for different types of faults: **a** F1, **b** F7, **c** F9, **d** F13, **e** F14, **f** F16

Table 3.4 Fault detection time(s) of the proposed method for 21 faults

Fault No.	F1	F2	F3	F4	F5	F6	F7
Time	0.96	0.99	1.09	0.78	1.04	0.91	0.95
Fault no.	F8	F9	F10	F11	F12	F13	F14
Time	0.97	0.94	1.01	0.74	1.00	0.94	0.91
Fault no.	F15	F16	F17	F18	F19	F20	F21
Time	0.98	0.95	0.78	1.02	0.99	0.93	0.76

CPU (3.4 GHz) and 8 GB RAM. As demonstrated in Table 3.4, the detection times are relatively short, with the maximum fault detection time being 1.09 s for F3 and the minimum being 0.76 s for F21. This discrepancy is attributed to the greater number of related variables in the subset of F3 compared to F21, implying that fewer related variables result in shorter detection times. Furthermore, the average detection time is 0.94 s. The complexity of the proposed method primarily arises from the kernel matrix. However, since KPCA is a shallow nonlinear modeling technique where the main computational task involves kernel model projection, and due to variable selection and block segmentation, the number of variables in each block is significantly reduced compared to the original number. Consequently, the computational and storage costs do not increase significantly. This confirms the method's effectiveness and computational efficiency.

The impact of varying window lengths on the monitoring performance of the proposed method is analyzed in this chapter. The window length is varied from 3 to 20. The monitoring results for different window lengths, along with the mean FDR and FAR, are presented in Fig. 3.9. From Fig. 3.9a, it can be observed that FDRs exhibit variations with the window length, though the changes are not significant. Similar observations can be made for most faults in Fig. 3.9b. However, FARs increase gradually with the window length when it exceeds 6, particularly for incipient faults and strongly coupled faults. This indicates that the FARs of a few faults are sensitive to the window length. Consequently, setting the window length to 6 appears to be a sensible choice. Additionally, the mean FDR and FAR for different window lengths are plotted in Fig. 3.9c. It is evident that the mean FDR varies considerably with window lengths from 3 to 6, and then changes gradually beyond a window length of 6. The mean FAR shows an increasing trend with the window length. Therefore, considering the computational cost and the trade-off between FDR and FAR, setting the window length to 6 is deemed reasonable.

3.4 Conclusion

In this chapter, we address the sample value imbalance problem in process monitoring. To enhance monitoring accuracy, we propose an improved SVDD method with a novel monitoring index called the restructured radius index. Initially, the KLD is employed as the variable selection algorithm to identify the most informative variables and eliminate irrelevant ones. The selected variables are grouped into blocks to ensure that relevant variables and critical information are concentrated within the same block. Within each block, KPCA is used to extract principal components, thus addressing the issues of high dimensionality and nonlinearity in the samples. The proposed SVDD method then generates the final monitoring results. Given that fault-dependent variables are selected and clustered within the same block, and the nonlinearity of the data is considered, the monitoring performance is significantly improved. Comparative results demonstrate that our proposed method exhibits superior performance for the TEP, achieving a mean FDR of 0.83. Specifically, it enhances the detection of incipient faults such as F3, F9, and F15, which are notoriously challenging to detect.

However, practical application of the proposed method is limited by the availability of fault samples, as typically only normal samples are easily collected. Future research will focus on integrating mechanistic knowledge to reduce dependence on

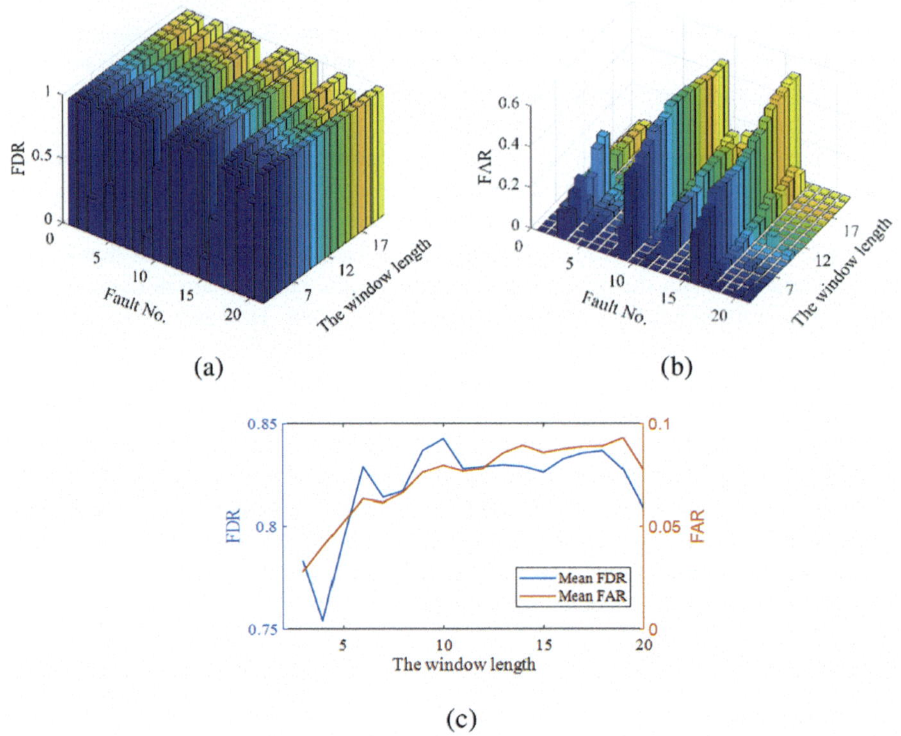

Fig. 3.9 The effect of the window length on monitoring result: **a** FDR, **b** FAR, **c** Mean FDR and FAR

fault samples, thereby enabling process monitoring with only one class of data and realizing more flexible and practical applications.

References

1. Liu C, Gryllias K (2020) A semi-supervised support vector data description-based fault detection method for rolling element bearings based on cyclic spectral analysis. Mech Syst Signal Process 140:106682
2. Zhao Y-P, Xie Y-L, Ye Z-F (2021) A new dynamic radius svdd for fault detection of aircraft engine. Eng Appl Artif Intell 100:104177
3. Chen K, Wang Z, Xiaowei G, Wang Z (2021) Multicondition operation fault detection for chillers based on global density-weighted support vector data description. Appl Soft Comput 112:107795
4. Scott D, Shang C, Huang B, Huang D (2020) A holistic probabilistic framework for monitoring nonstationary dynamic industrial processes. IEEE Trans Control Syst Technol 29(5):2239–2246
5. Jiang Q, Chen S, Yan X, Kano M, Huang B (2021) Data-driven communication efficient distributed monitoring for multiunit industrial plant-wide processes. IEEE Trans Autom Scie Eng 19(3):1913–1923
6. Huang J, Yang X, Peng K (2020) Double-layer distributed monitoring based on sequential correlation information for large-scale industrial processes in dynamic and static states. IEEE Trans Indus Inform 17(9):6419–6428
7. Jiang Q, Yan X, Huang B (2015) Performance-driven distributed pca process monitoring based on fault-relevant variable selection and bayesian inference. IEEE Trans Indus Electron 63(1):377–386
8. Zhu J, Shi H, Song B, Tao Y, Tan S (2021) Convolutional neural network based feature learning for large-scale quality-related process monitoring. IEEE Trans Indus Inform 18(7):4555–4565
9. Zhou B, Xingsheng G (2020) Multi-block statistics local kernel principal component analysis algorithm and its application in nonlinear process fault detection. Neurocomputing 376:222–231

Chapter 4
Industrial Process Monitoring Using Fisher Discriminant-Based Global-Local Preserving Projection

A novel data-driven approach for process fault detection, called Fisher Discriminant Global-Local Preserving Projection (FDGLPP), is introduced. This method combines Fisher Discriminant Analysis with Global-Local Preserving Projection to determine the optimal projection direction, then transforms the original data using the projection matrix. This approach preserves the dataset's global manifold structure, local neighborhood relations, and the discriminative power of the projection subspace, thereby enhancing fault identification accuracy. Additionally, to mitigate the impact of noise and measurement errors, Kernel Density Estimation (KDE) is used to calculate the control limits for monitoring indices, which reduces false alarms and improves detection efficiency, enhancing monitoring effectiveness. Numerical simulations validate the dimensionality reduction advantages of the FDGLPP algorithm, and a case study on the Tennessee Eastman process demonstrates that this method significantly outperforms GLPP, LPP, and PCA in fault detection and identification. Finally, t-SNE is used to visualize FDGLPP's discriminative capability.

4.1 Motivation

With advancements in sensor technology and computing power, MSPM techniques are increasingly used to enhance industrial operation quality and safety [1]. These data-driven methods leverage historical data and process insights to determine optimal projection directions, extracting feature variables that effectively represent the relationships between parameters and the system's operational state. Process monitoring typically involves four stages: fault detection, fault diagnosis, fault isolation, and fault evaluation. Fault detection identifies abnormal conditions or failures, while fault diagnosis classifies specific fault types using supervised techniques [2]. Fault isolation pinpoints root causes when process information is limited, and fault evaluation assesses fault severity, offering insights to prioritize responses.

H. Yin et al., *Data-Driven Fault Diagnosis for Complex Industrial Processes*,
Engineering Applications of Computational Methods 22,
https://doi.org/10.1007/978-981-96-3153-7_4

Traditional projection methods like PCA, partial least squares (PLS), and local preserving projection (LPP) are commonly used for dimensionality reduction and fault detection [3]. PCA, widely applied in process monitoring, decomposes data into a principal component subspace, capturing normal process variations, and a residual subspace, where fault information may reside. However, it retains only the global structure, neglecting local features related to fault types and causes. In contrast, LPP emphasizes local structure but overlooks the global context. These methods assume data and statistics (e.g., SPE and Hotelling's T^2) follow a Gaussian distribution, but noise and measurement errors in real systems disrupt this assumption, reducing diagnostic accuracy [4]. To enhance accuracy, this chapter applies Kernel Density Estimation (KDE) to calculate control limits for SPE and T^2 statistics.

GLPP preserves both global and local structures of data, offering more flexibility than PCA and LPP [5]. To handle nonlinear relationships, Kernel GLPP (KGLPP) was introduced for fault detection [6]. However, traditional GLPP methods often capture only part of the global-local structure. To improve this, Zhan developed Ensemble GLPP (EGLPP) by integrating ensemble learning and Bayesian inference, enhancing detection performance while reducing false alarms [7]. Additionally, Bao and colleagues proposed Sparse Global-Local Preserving Projection (SGLPP), which provides more interpretable results by revealing underlying process mechanisms and fault propagation across variables [8].

Fisher Discriminant Analysis (FDA) excels in identifying discriminative directions for fault diagnosis, outperforming PCA- and PLS-based methods focused on data representation [9]. While methods like SGLPP, EGLPP, and KGLPP prioritize reconstruction directions, the proposed Fisher Discriminant Global-Local Preserving Projection (FDGLPP) enhances fault type classification by finding optimal discriminative directions. FDGLPP addresses FDA's limitations with sample overlap by incorporating both Euclidean and manifold structures of monitoring data [10].

This chapter introduces a novel fault detection and identification method, called FDGLPP, along with two-level deviation magnitude plots for industrial systems. FDGLPP combines Fisher Discriminant Analysis with the GLPP method by incorporating additional constraints within the GLPP framework to maximize inter-class scatter and minimize intra-class scatter. This approach not only preserves the local neighborhood structure and global manifold characteristics of the data but also improves the interpretability and accuracy of the projection directions. By integrating Fisher Discriminant Analysis into the GLPP model, the method enhances class separation and alleviates the sample overlap issues commonly encountered in FDA. Furthermore, KDE is employed to set control limits for the statistics, mitigating the impact of noise and measurement errors in industrial environments. These adjustments make FDGLPP more adaptable and effective for practical fault diagnosis. The performance of the FDGLPP algorithm is validated through numerical simulations and a case study on the TEP.

4.2 Methodology

4.2.1 Revised of Fisher Discriminant Analysis and GLPP

FDA is a linear technique widely used in pattern classification and dimensionality reduction [9]. This method projects the original dataset onto a discriminative direction by maximizing inter-class scatter and minimizing intra-class scatter, thereby achieving better separation of samples from different classes.

Let the data sets be denoted as $\mathbf{X}_l \in \mathbb{R}^{m \times n_l}, l = 0, 1, \ldots, c$, let $\mathbf{X}_0 \in \mathbb{R}^{m \times n_0}$ represents the normal data set, scaled to have zero mean and unit variance, with n_0 samples and m dimensions. The fault data sets $\mathbf{X}_k \in \mathbb{R}^{m \times n_k}, k = 1, 2, \ldots, c$, are scaled to have the same mean and variance as the normal data set, where c represents the number of fault classes and n_k is the number of samples in class k. The overall fault data matrix is given by $\mathbf{X} = \cup_{k=1}^{c} \mathbf{X}_k$, and each column $\mathbf{x}_i$ of $\mathbf{X}$ represents an individual sample. FDA seeks to find a set of optimal projection vectors $\mathbf{a}_i \in \mathbb{R}^{m \times 1}$ by minimizing the within-class scatter and maximizing the between-class scatter simultaneously. Stacking all the projection vectors $\mathbf{a}_i$ together forms the Fisher transformation matrix $\mathbf{A} \in \mathbb{R}^{m \times d}, d < m$, where d is the dimensionality of the projected subspace, with $d < m$. Each data sample $\mathbf{x}_i$ is then projected onto the lower-dimensional subspace as follows:$\mathbf{y}_i = \mathbf{A}^T \mathbf{x}_i$, where $\mathbf{y} \in \mathbb{R}$ is the projected data sample in the lower-dimensional subspace as:

$$\mathbf{y}_i = \mathbf{A}^T \mathbf{x}_i \tag{4.1}$$

$$\mathbf{Y} = \mathbf{A}^T \mathbf{X} \tag{4.2}$$

Denote the within class scatter matrix $\mathbf{S}_w$ and between classes scatter matrix $\mathbf{S}_b$ of $\mathbf{X}$ as:

$$\mathbf{S}_b = \sum_{k=1}^{c} n_k (\mathbf{m}_k - \mathbf{m})(\mathbf{m}_k - \mathbf{m})^T \tag{4.3}$$

$$\mathbf{S}_w = \sum_{k=1}^{c} \sum_{\mathbf{y}_i \in \mathbf{Y}_k} (\mathbf{y}_i - \mathbf{m}_k)(\mathbf{y}_i - \mathbf{m}_k)^T \tag{4.4}$$

where $\mathbf{m}_k = \frac{1}{n_k} \sum_{i=1}^{n_k} \mathbf{y}_i$ is the mean vector of all samples in class k, $\mathbf{m} = \frac{1}{c} \sum_{i=1}^{c} \mathbf{m}_i$ is the mean vector of all samples in dataset $\mathbf{Y}$.

Intuitively, the optimal Fisher projected vectors can be acquired by solve the objective function of FDA as [11]:

$$J_{Fisher} = \arg\min_{\mathbf{a}} \{tr(\mathbf{S}_w - \mathbf{S}_b)\} \tag{4.5}$$

The GLPP (Global-Local Preserving Projection) algorithm is designed to construct a projection subspace that preserves both the local neighborhood structure and the global manifold structure of the data [12]. Let the local neighborhood weight matrix be denoted by **W** and the global weight matrix by $\bar{W}$. The adjacency relationship between data points $\mathbf{x}_i$ and $\mathbf{x}_j$ is represented by the i_{th}, j_{th} element W_{ij} of **W**, and the non-adjacency relationship between $\mathbf{x}_i$ and $\mathbf{x}_j$ is represented by the i_{th}, j_{th} element $\bar{W}_{ij}$ of $\bar{W}$. GLPP aims to find a set of projection vectors $\mathbf{a}_i \in \mathbb{R}^{d\times 1}$, which together form the transformation matrix $\mathbf{A} \in \mathbb{R}^{m\times d}$, $d < m$, to map the original data set to a lower-dimensional subspace. The local and global weights w_{ij} and $\bar{w}_{ij}$ are defined as follows:

$$w_{ij} = \begin{cases} e^{-\frac{\|\mathbf{x}_i-\mathbf{x}_j\|^2}{\sigma_1}}, if\ \mathbf{x}_j \in \Omega_k(\mathbf{x}_i) \text{ or } \mathbf{x}_i \in \Omega_k(\mathbf{x}_j) \\ 0, \text{otherwise} \end{cases} \tag{4.6}$$

$$\bar{w}_{ij} = \begin{cases} e^{-\frac{\|\mathbf{x}_i-\mathbf{x}_j\|^2}{\sigma_2}}, if\ \mathbf{x}_j \notin \Omega_k(\mathbf{x}_i) \text{ and } \mathbf{x}_i \notin \Omega_k(\mathbf{x}_j) \\ 0, \text{otherwise} \end{cases} \tag{4.7}$$

where σ_1 and σ_2 is the constant parameters, and $\Omega_k(\mathbf{x}_i)$ is the neighborhood of data sample $\mathbf{x}_i$, which is defined by k nearest neighbors.

The objective function can be described as:

$$J_{GLPP} = \arg\min_{\mathbf{a}} \frac{1}{2}\left\{\sum\nolimits_{i,j}\left(\eta(\mathbf{y}_i-\mathbf{y}_j)^2 W_{ij} - (1-\eta)\sum\nolimits_{i,j}(\mathbf{y}_i-\mathbf{y}_j)^2 W_{ij}\right)\right\} \tag{4.8}$$

where $\mathbf{y}_i$ is the projection of $\mathbf{x}_i$ by $\mathbf{y}_i = \mathbf{A}^T\mathbf{x}_i$, Eq. (4.8) can be rewritten as:

$$\begin{aligned} J_{GLPP} &= \arg\min_{\mathbf{a}} \frac{1}{2}\left\{\sum\nolimits_{i,j}\left(\eta(\mathbf{y}_i-\mathbf{y}_j)^2 W_{ij} - (1-\eta)\sum\nolimits_{i,j}(\mathbf{y}_i-\mathbf{y}_j)^2 \bar{W}_{ij}\right)\right\} \\ &= \arg\min_{\mathbf{a}} \frac{1}{2}\left\{\sum\nolimits_{i,j}\left((\mathbf{y}_i-\mathbf{y}_j)^2 R_{ij}\right)\right\} \\ &= \arg\min_{\mathbf{a}}\left\{\sum\nolimits_{i}\left(\mathbf{a}^T\mathbf{x_i}\mathbf{H_{ii}}\mathbf{x}_i^T\mathbf{a}\right) - \sum\nolimits_{i,j}\left(\mathbf{a}^T\mathbf{x}_i\mathbf{R}_{ij}\mathbf{x}_j^T\mathbf{a}\right)\right\} \\ &= \arg\min_{\mathbf{a}}\left\{a^T\mathbf{X}(\mathbf{H}-\mathbf{R})\mathbf{X}^T a\right\} \\ &= \arg\min_{\mathbf{a}}\left\{a^T\mathbf{XMX}^T a\right\} \end{aligned} \tag{4.9}$$

where $R_{ij} = \eta W_{ij} - (1-\eta)\bar{W}_{ij}$, **H** is a diagonal matrix consists of $H_{ii} = \sum_j R_{ij}$, $\mathbf{M} = \mathbf{H} - \mathbf{W}$ is the Laplacian matrix, η is a constant to balance the local and global weight. Furthermore, a constraints is imposed on GLPP for normalization, then the objective function of GLPP is defined as follows:

$$J_{GLPP} = \arg\min_{\mathbf{a}}\left\{\mathbf{a^T XMX^T a}\right\} \quad \text{s.t.} \quad \mathbf{a^T XNX^T a} = 1 \tag{4.10}$$

where $\mathbf{N} = \eta \mathbf{X}\mathbf{H}\mathbf{X}^T + (1-\eta)\,\mathbf{I}_m$.

Then this optimization problem is transferred to a generalized eigenvalue problem as:

$$\mathbf{X}\mathbf{M}\mathbf{X}^T\mathbf{a} = \lambda \mathbf{X}\mathbf{N}\mathbf{X}^T\mathbf{a} \tag{4.11}$$

The optimal LPP projection matrix $\mathbf{A}$ is consist of the eigenvectors $a_1, a_2, ..., a_l$, the sequence is sorted by the corresponding eigenvalues $\lambda_1 < \lambda_2 <, ..., < \lambda_l$.

4.2.2 Formulation of FDGLPP

Considering that GLPP is effective in capturing the data's representation while FDA excels in discrimination, the approach integrates GLPP with FDA into a FDGLPP. This method projects the original data into a more discriminative subspace, improving dimensionality reduction, especially when dealing with overlapping datasets. To combine FDA with GLPP, Eqs. (4.3) and (4.4) can be reformulated as follows:

$$\mathbf{S}_b = \frac{1}{2}\sum\nolimits_{i,j}\left(\mathbf{y}_i - \mathbf{y}_j\right)^2 W_{ij}^b \tag{4.12}$$

$$\mathbf{S}_w = \frac{1}{2}\sum\nolimits_{i,j}\left(\mathbf{y}_i - \mathbf{y}_j\right)^2 W_{ij}^w \tag{4.13}$$

where W_{ij}^b is the $i\,j_{th}$ element of between classes scatter matrix $\mathbf{W}^b$, and W_{ij}^w is the $i\,j_{th}$ element of within class scatter matrix $\mathbf{W}^w$, $\mathbf{W}^b$ and $\mathbf{W}^w$ is defined as:

$$w_{ij}^b = \begin{cases} \frac{1}{n_l} - \frac{1}{n_k}, if\ \mathbf{x}_j \in \mathbf{X}_i \text{ and } \mathbf{x}_i \in \mathbf{X}_j \\ \frac{1}{n_l}, \text{ otherwise} \end{cases} \tag{4.14}$$

$$w_{ij}^w = \begin{cases} \frac{1}{n_k}, if\ \mathbf{x}_j \in \mathbf{X}_i \text{ and } \mathbf{x}_i \in \mathbf{X}_i \\ 0, \text{ otherwise} \end{cases} \tag{4.15}$$

According to Eqs. (4.14) and (4.15), $\mathbf{S}_b$ increases as the distance between samples from different classes grows, while it does not affect $\mathbf{S}_w$. Conversely, $\mathbf{S}_b$ decreases as the distance between samples from the same class increases, whereas $\mathbf{S}_w$ increases.

To integrate FDA and GLPP, the objective function of FDA can be rewritten in a more general form as follows:

$$J_{Fisher} = \arg\min_{\mathbf{a}} \{\eta \mathbf{S}_w - (1-\eta)\,\mathbf{S}_b\} \tag{4.16}$$

where η is a constant, Eq. (4.5) is the special form of Eq. (4.16) while $\eta = \frac{1}{2}$.

Then the Eq. (4.5) can be rewritten as:

$$J_{Fisher} = \arg\min_{\mathbf{a}} \left\{ \frac{1}{2} \sum\nolimits_{i,j} \left(\eta (\mathbf{y}_i - \mathbf{y}_j)^2 W_{ij}^b - (1-\eta) \sum\nolimits_{i,j} (\mathbf{y}_i - \mathbf{y}_j)^2 W_{ij}^w \right) \right\} \tag{4.17}$$

It is clearly that the Eq. (4.17) has the same form with Eq. (4.8), then the Fisher discriminant global and local preserving projection is defined as:

$$\begin{aligned} J_{FDGLPP} &= J_{GLPP} + J_{Fisher} \\ &= \arg\min_{\mathbf{a}} \frac{1}{2} \left\{ \sum\nolimits_{i,j} \left(\eta (\mathbf{y}_i - \mathbf{y}_j)^2 W_{ij} - (1-\eta) \sum\nolimits_{i,j} (\mathbf{y}_i - \mathbf{y}_j)^2 \bar{W}_{ij} \right) \right\} \\ &+ \arg\min_{\mathbf{a}_i} \left\{ \frac{1}{2} \sum\nolimits_{i,j} \left(\eta (\mathbf{y}_i - \mathbf{y}_j)^2 W_{ij}^b - (1-\eta) \sum\nolimits_{i,j} (\mathbf{y}_i - \mathbf{y}_j)^2 W_{ij}^w \right) \right\} \\ &= \arg\min_{\mathbf{a}} \frac{1}{2} \left\{ \sum\nolimits_{i,j} \left(\eta (\mathbf{y}_i - \mathbf{y}_j)^2 \left(W_{ij} + W_{ij}^b \right) - (1-\eta) \sum\nolimits_{i,j} (\mathbf{y}_i - \mathbf{y}_j)^2 \left(\bar{W}_{ij} + W_{ij}^w \right) \right) \right\} \\ &= \arg\min_{\mathbf{a}} \frac{1}{2} \left\{ \sum\nolimits_{i,j} \left(\eta (\mathbf{y}_i - \mathbf{y}_j)^2 \tilde{W}_{ij} - (1-\eta) \sum\nolimits_{i,j} (\mathbf{y}_i - \mathbf{y}_j)^2 \hat{W}_{ij} \right) \right\} \end{aligned} \tag{4.18}$$

where $\tilde{W} = W_{ij} + W_{ij}^b$, $\hat{W} = \bar{W}_{ij} + W_{ij}^w$.

Imitate the transformation of the GLPP model as Eqs. (4.9), (4.18) is rewritten as:

$$\begin{aligned} J_{FDGLPP} &= J_{GLPP} + J_{Fisher} \\ &= \arg\min_{\mathbf{a}} \frac{1}{2} \left\{ \sum\nolimits_{i,j} \left(\eta (\mathbf{y}_i - \mathbf{y}_j)^2 \tilde{W}_{ij} - (1-\eta) \sum\nolimits_{i,j} (\mathbf{y}_i - \mathbf{y}_j)^2 \hat{W}_{ij} \right) \right\} \\ &= \arg\min_{\mathbf{a}} \frac{1}{2} \left\{ \sum\nolimits_{i,j} \left((\mathbf{y}_i - \mathbf{y}_j)^2 \tilde{R}_{ij} \right) \right\} \\ &= \arg\min_{\mathbf{a}} \left\{ \sum\nolimits_{i} \left(a^T x_i \tilde{H}_{ii} x_i^T a \right) - \sum\nolimits_{i,j} \left(a^T x_i \tilde{R}_{ij} x_j^T a \right) \right\} \\ &= \arg\min_{\mathbf{a}} \left\{ a^T \mathbf{X} \left(\tilde{H} - \tilde{R} \right) \mathbf{X}^T a \right\} \\ &= \arg\min_{\mathbf{a}} \left\{ a^T X \tilde{M} \mathbf{X}^T a \right\} \end{aligned} \tag{4.19}$$

where $\tilde{R}_{ij} = \eta \tilde{W}_{ij} - (1-\eta) \hat{W}_{ij}$, $\mathbf{H}$ is a diagonal matrix consists of $\tilde{H}_{ii} = \sum_j \tilde{R}_{ij}$, $\mathbf{M} = \mathbf{H} - \mathbf{W}$ is the Laplacian matrix.

4.2.3 Process Monitoring Based on Fisher Discriminant GLPP

Let $\mathbf{X} \in \mathbb{R}^{n \times m}$ be the data matrix, where n represents the number of samples and m is the number of dimensions. The data matrix $\mathbf{X}$ is scaled to have zero mean and unit variance, with the normal dataset consisting of n samples and m dimensions. FDGLPP aims to find a set of projection vectors $\mathbf{a}_i \in \mathbb{R}^m$ that map the data samples into a new space.Let the transformation matrix be denoted as $\mathbf{A} = [\mathbf{a_1}, \mathbf{a_2}, \ldots, \mathbf{a_d}]$, where d is the dimensionality of the new subspace, and each $\mathbf{a}_i \in \mathbb{R}^m$ is a projection vector. Then, a data sample $\mathbf{x}_i \in \mathbb{R}^m$ is projected into the lower-dimensional subspace as:$\mathbf{y}_i \in \mathbb{R}^d$

$$\mathbf{y}_i = \mathbf{A}^T \mathbf{x}_i \tag{4.20}$$

$$\mathbf{x}_i = \left(\mathbf{A}^T\right)^+ \mathbf{y}_i + \mathbf{e}_i \tag{4.21}$$

$$\mathbf{A} = \left(\mathbf{A}^T\right)^+ \mathbf{Y} + \mathbf{E} \tag{4.22}$$

The T^2 and SPE statistics are introduced for process state monitoring. The T^2 index quantifies the variation magnitude within the principal component subspace, while the SPE index captures variability that deviates from normal process correlations, typically indicating abnormal or faulty conditions. Since SPE is defined in the residual subspace, it is generally more sensitive to fault detection.For a sample $\mathbf{x}_i$, the definitions of T^2 and SPE are as follows:

$$T^2 = \mathbf{y}_i^T \mathbf{S}^{-1} \mathbf{y}_i \tag{4.23}$$

$$SPE = \mathbf{e}_i^T \mathbf{e}_i = \left(\mathbf{x}_i - \left(\mathbf{A}^T\right)^+ \mathbf{y}_i\right)^2 \tag{4.24}$$

where $S = \mathbf{Y}^T \mathbf{Y} / (n - 1)$ is the covariance matrix of the projection matrix $\mathbf{Y}$.

To determine the threshold for T^2 and SPE, it is generally assumed that the T^2 statistic follows a weighted F-distribution, and the SPE statistic adheres to a weighted χ^2-distribution. However, in practical applications, factors like noise and measurement inaccuracies disrupt these expected distributions, leading to a decrease in both the precision and responsiveness of process monitoring. To address this, KDE is utilized in this chapter to more effectively calculate the thresholds.

Define the T^2 of all training samples $\mathbf{X} = [\mathbf{x}_1, \mathbf{x}_2, \ldots \mathbf{x}_n]$ is $\left[T_1^2, T_2^2, \ldots T_n^2\right]$, then the probability density function of T^2 is :

$$\hat{f}(t) = \frac{1}{nh} \sum_{i=1}^{n} \left(\frac{t - T_i^2}{h}\right) \tag{4.25}$$

Here, the kernel function is denoted as n represents the number of data samples, and h is the window width, also known as the bandwidth or smoothing parameter. The kernel function defines the shape of the bumps, while the parameter h controls their width.

This chapter selects the Gaussian kernel function. There are various methods for determining the value of h, such as cross-validation and plug-in approaches. Mugdadi and Ahmad [13] have conducted research on selecting h, and they employ the least-squares cross-validation method to determine an appropriate value. The parameter h can be computed as follows:

$$h = 1.06\sigma n^{-0.2} \tag{4.26}$$

Here, σ represents the standard deviation of the data sample. The cumulative distribution function $P\,(x < \theta)$ can be derived from the probability density function $\hat{f}\,(t)$ as follows:

$$P\,(x < \theta) = \int_{-\infty}^{\theta} f\,(x)dx = 1 - \alpha \tag{4.27}$$

To determine the confidence limit for the T^2 statistic at a significance level α (where $\alpha = 0.05$ or $\alpha = 0.05$), the control limit is calculated by evaluating the upper bound of the integral in Eq. (4.2.3) [7]. A similar approach is used to compute the control limit for the SPE index. Fault detection occurs when either the T^2 or SPE index of a new sample exceeds its corresponding control limit.

4.2.4 *The Overall Framework*

The overall framework is illustrated in Fig. 4.1.

Off-line Training Process

- Use monitoring data from the normal operating state as training data and normalize it.
- Calculate the local weight coefficients w_{ij} and global weight coefficients $\tilde{w}_{ij}$ for $\mathbf{x}_i$ and $\mathbf{x}_j$ in the training data.

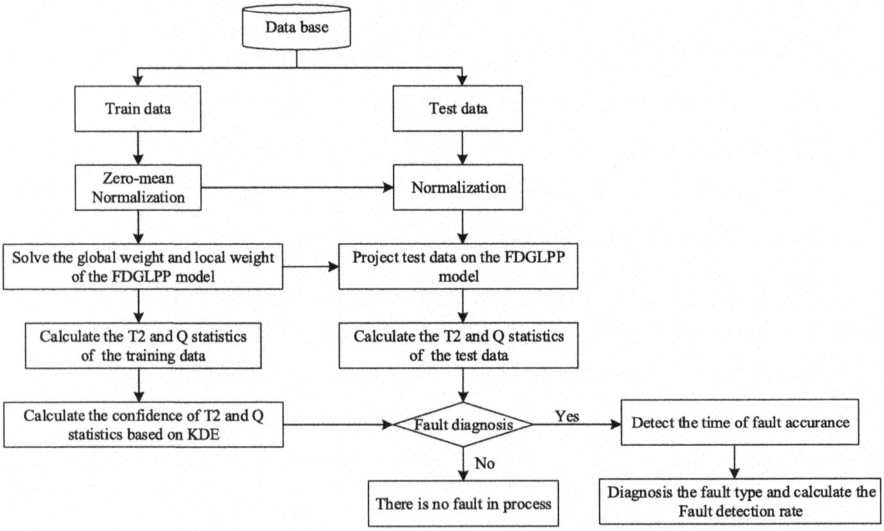

Fig. 4.1 Monitoring procedure based on FDGLPP algorithm

- Solve for the projection matrix using the FDGLPP model.
- Project the training data onto the projection matrix and calculate its T^2 and SPE statistics.
- Determine the confidence limits of the T^2 and SPE statistics through Kernel Density Estimation (KDE).

Online Monitoring Process

- Treat the real-time monitoring data as the test set and standardize it using the mean and standard deviation from the training set.
- Project the test data onto the projection matrix generated during the offline phase to achieve dimensionality reduction.
- Calculate the real-time T^2 and SPE statistics of the test data and compare them with the confidence limits to determine if any anomalies exist.
- If an anomaly is detected, identify the time of occurrence, classify the fault type, and calculate the detection rate.

4.3 Computing Complexity

Given the high computational complexity, the FDGLPP algorithm's complexity is divided into three main components: finding the K nearest neighbors, solving the eigenvector problem, and calculating the control limit using KDE. Let the training data be $X \in \mathbb{R}^{m \times n}$, where m represents the dimensions and n the number of samples. $Y \in \mathbb{R}^{d \times n}$ is the dataset in the feature subspace obtained by projection through the transformation matrix, and k denotes the number of neighbors required in the search process.

Firstly, the complexity of searching the K nearest neighbors is calculated as follows:

$$T_{KNN} = O\left((m+d)\,n^2\right) \tag{4.28}$$

Secondly, resolving the eigenvector problem occurs in finding the FDGLPP transformation vectors, the complexity is calculated as follows:

$$T_{eig} = O\left(dm^2\right) \tag{4.29}$$

Thirdly, complexity of calculating the control limit by KDE is calculated as follows:

$$T_{KDE} = O\left(n^2\right) \tag{4.30}$$

As a result, the complexity of FDGLPP algorithm is defined as:

$$\begin{aligned} T_{FDGLPP} &= T_{KNN} + T_{eig} + T_{KDE} \\ &= O\left((m+d)\,n^2\right) + O\left(dm^2\right) + O\left(n^2\right) \\ &= O\left((m+d+1)\,n^2 + dm^2\right) \end{aligned} \tag{4.31}$$

4.4 Case Study

4.4.1 Numerical Simulation

This research employed two synthetic datasets—the S-curve and Swiss roll–which are two-dimensional manifolds embedded in three-dimensional space, to showcase the dimensionality reduction performance of the proposed algorithm. Each dataset consists of 2000 points, with the neighborhood size set to $k = 10$ and the balance parameter for GLPP and FDGLPP set to $\beta = 0.6$. Additionally, PCA and GLPP were included as benchmark methods for comparison with FDGLPP.

Figures 4.2 and 4.3 show the dimension reduction results of the S-curve and Swiss roll datasets using different methods, respectively. It is evident that in the PCA-generated illustrations, the data points are mixed, as seen in Figs. 4.2b and 4.3b. This is because PCA retains only the global structure by maximizing variance, ignoring the neighborhood relationships between data points. Additionally, PCA's dimensionality reduction is suited to data representation rather than pattern recognition. GLPP, on the other hand, performs better, as shown in Figs. 4.2c and 4.3c, retaining both the global structure and some neighborhood relationships, though data points remain challenging to classify. Figures 4.2d and 4.3d demonstrate the notable advantages of the FDGLPP algorithm; by introducing discriminant constraints, the dimensionality reduction achieves clear separation while preserving both the global and local structures of the original data.

4.4.2 Tennessee Eastman Process

The FDGLPP algorithm is applied to the TEP. In this chapter, normal operational training samples are used to train the FDGLPP model, and its effectiveness is verified by applying it to an additional 21 fault test datasets. To more comprehensively demonstrate the performance of this method, sparse exponential discriminant analysis (SEDA) [2], PCA [3], LPP [14], and GLPP [7] are included as comparative experiments. The balance parameter for FDGLPP is set to $\eta = 0.6$, with constant parameters for global and local weight calculations set at $\sigma_1 = 48$ and $\sigma_2 = 132$, respectively. The neighborhood number k for each data sample bfx_i is set to 10, and the significance level for the FDGLPP model monitoring statistics is set to $\alpha = 0.01$.

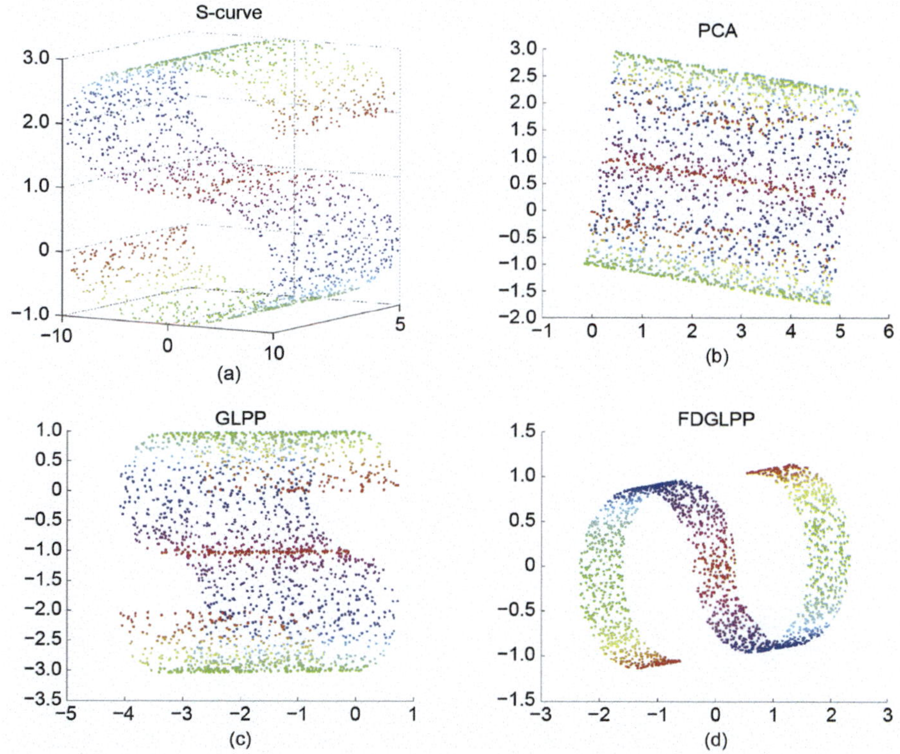

Fig. 4.2 Dimension reduction results of S-curve by different methods

FDR, detection delay, and false alarm are used as performance evaluation metrics for diagnosis. Detection delay refers to the time interval between when a fault occurs and when it is detected. False alarm indicates the number of normal condition samples that are incorrectly identified as being in a fault state. The FDR is defined based on classification accuracy as follows:

$$
\begin{aligned}
FDR\,(i) &= \frac{n_i^T}{n_i} \times 100\% \\
anvFDR &= \frac{1}{K} \times \sum_{i=1}^{K} FDR\,(i)
\end{aligned}
\tag{4.32}
$$

where $FDR\,(i)$ is the classification accuracy of the i_{th} class, n_i^T and n_i is the number of the correct classified samples and the total number of samples in the i_{th} class respectively. $anvFDR$ is the average classification accuracy of all kinds of faults, K is the number of fault types.

To evaluate the fault diagnosis performance of the proposed FDGLPP method against SEDA, PCA, LPP, and GLPP, all 21 faults in the TEP were tested, and the FDR results are presented in Table 4.1. It is evident that FDGLPP achieves the

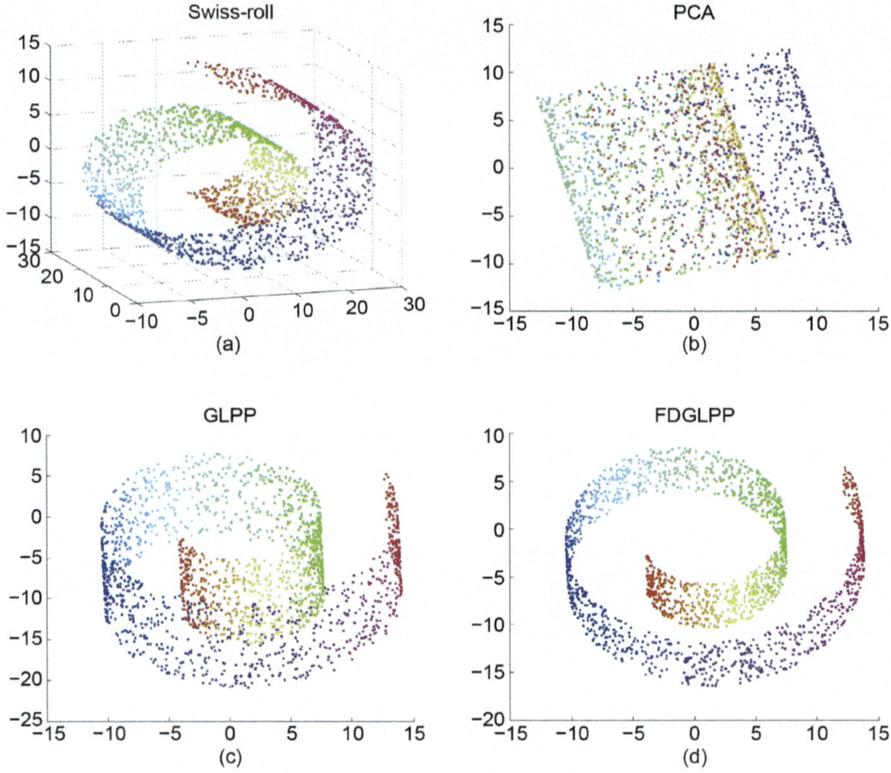

Fig. 4.3 Dimension reduction results of Swiss roll by different methods

highest average detection rate, while SEDA records the lowest. FDGLPP accurately diagnoses faults 1, 2, 5, 6, 7, 13, 14, and 18 with nearly 100% accuracy and shows significant improvement in diagnosing faults 10, 16, 17, and 20 compared to the GLPP method. This demonstrates that the FDGLPP algorithm is more sensitive to faults of smaller magnitude, and the integration of Fisher discriminant limitations with GLPP greatly enhances the discrimination within the projection space. Faults 3 and 9 are closely related to changes in the condenser cooling water inlet temperature; when this temperature fluctuates suddenly, the closed control loop adjusts to maintain system balance, which can eliminate fault features and complicate diagnosis. This aligns with the lowest detection rates for faults 3 and 9 across all methods shown in Table 4.1.

As illustrated in the T^2 graphs in Fig. 4.4a, b, the FDGLPP method diagnoses fault 5 without any detection delay or false alarms among the 160 samples from the initial process run. In contrast, the GLPP method encounters multiple false alarms during the monitoring process. After a fault is detected, the monitoring values obtained through the FDGLPP algorithm consistently exceed the control level of the Squared Prediction Error (SPE), whereas almost all samples after the 250th point cannot be

Table 4.1 Comparison of fault detection rates of SEDA, PCA, LPP, GLPP, FDGLPP

Fault/FDR	SEDA	PCA	LPP	GLPP	FDGLPP
Fault1	81.46	97.88	99.13	99.25	99.34
Fault2	80.42	96.5	98	98.25	98.25
Fault3	11.46	11.25	15	11.38	27.75
Fault4	20.21	14.88	76.88	81.75	89.75
Fault5	35.63	24	87.5	81.25	99.38
Fault6	77.19	99.5	100	100	100
Fault7	76.98	90.75	90.38	81.37	100
Fault8	29.27	26.88	56.13	96.63	67.75
Fault9	2.83	10.5	14.75	1.25	17.25
Fault10	10.32	13.5	21.88	11.75	66.5
Fault11	20.65	20.63	39.88	28.63	46.88
Fault12	20.21	40.5	44.13	93.5	91.88
Fault13	25.62	25.75	33.5	92.37	95.13
Fault14	65	77	95.63	86.62	95.88
Fault15	13.44	13.38	15.88	14	17.88
Fault16	4.77	9.13	44.25	6.63	83.13
Fault17	46.98	59.25	79.5	64.5	93.75
Fault18	62.4	83	84.38	88.75	97.38
Fault19	21.67	49.38	66.13	68.25	87.5
Fault20	34.79	21.63	34.25	29.38	59.38
Fault21	15.2	10.125	12.5	28	44.75
Average	36.02	42.64	57.61	60.17	75.21

detected by the GLPP method. Table 4.1 indicates that the fault detection rate for fault 5 using FDGLPP is significantly higher than that of GLPP, which aligns with the monitoring performance shown in Fig. 4.4a, b.

When comparing Fig. 4.4c, d, both methods detect faults without detection delays or false alarms. However, most monitoring values from the GLPP samples after the 323th point fail to be identified by the control limits, while FDGLPP continues to trigger alarms during process monitoring after a fault is detected.

In summary, the GLPP algorithm not only produces false alarms but also experiences a loss of effectiveness in its control limits after a fault is detected, leading to suboptimal monitoring performance. In contrast, the FDGLPP algorithm is capable of monitoring processes with minimal false alarms while maintaining continuous alarms throughout the entire procedure after a fault is identified.

As shown in the T^2 graphs in Fig. 4.5a, b, the FDGLPP algorithm successfully detects fault 13 at the 196th sample without any false alarms and maintains continuous alarms after the 207th point. In contrast, the GLPP method has a detection delay of 40 samples, triggering alarms only after the 210th point. Additionally, the GLPP method

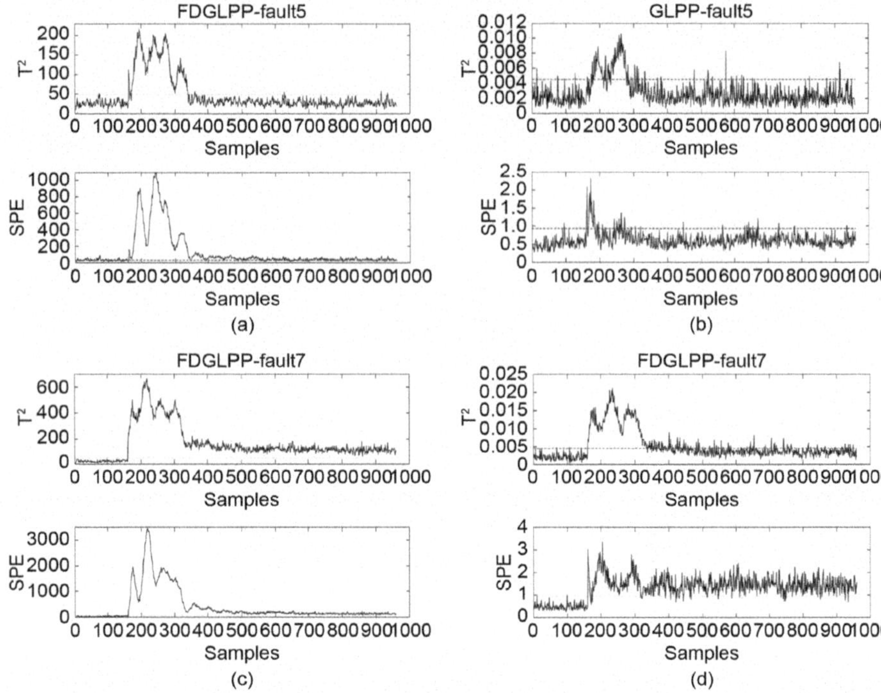

Fig. 4.4 Process monitoring of fault 5 and fault 7 by GLPP and FDGLPP algorithm

loses its control limits between the 305th and 340th points. Comparing Fig. 4.5a, b, the FDGLPP algorithm accurately identifies fault 17 at the 188th point without any errors and continues to sound alarms, whereas the GLPP method can only detect the fault after the 200th point. The Squared Prediction Error (SPE) index shows superior performance in detecting fault 17, identifying it simultaneously with the FDGLPP algorithm and maintaining alarms after the 180th point, which is 25 points earlier than the GLPP method. Overall, the FDGLPP algorithm demonstrates greater sensitivity to faults during process monitoring and has a shorter detection delay compared to GLPP.

The main goal of fault diagnosis is to accurately identify the type of fault, which relies heavily on supervised classification methods. By using the Fisher discriminant global-local preserving projection (FDGLPP) algorithm to determine the optimal projection matrix, the transformed training data can be used to assess the classification performance of the techniques. To evaluate the effectiveness of the proposed FDGLPP algorithm, t-SNE (t-distributed stochastic neighbor embedding) is applied to visualize the diagnostic outcomes of FDGLPP, GLPP, and LPP. t-SNE provides a way to represent high-dimensional data intuitively by mapping each sample into a two- or three-dimensional space, enabling a clear visualization of the relationships between neighboring samples [15].

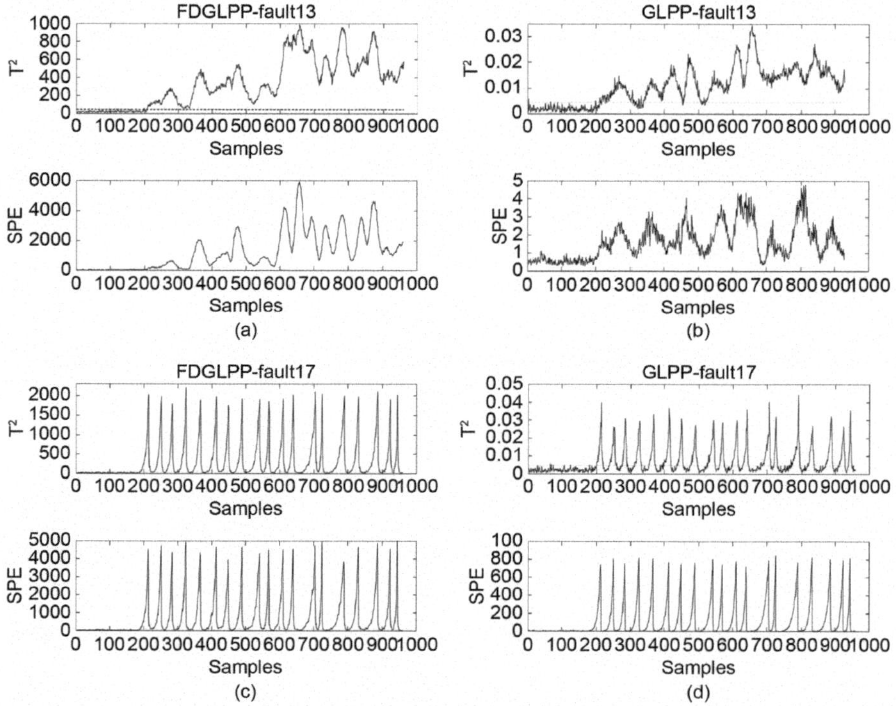

Fig. 4.5 Process monitoring of fault 13 and fault 17 by GLPP and FDGLPP algorithm

The experiment selected faults 1, 2, 5, 6, 7, 13, 15, and 18 to comprehensively assess the efficiency of the proposed algorithm. Additionally, it included step, slow drift, sticking, and unknown fault types. The projections of these eight types of fault data were reduced to three-dimensional space, as shown in Fig. 4.6. Figure 4.6a, c, e present the 3D views at an azimuth of $-37.5°$ and a top view of $30°$. To better illustrate the classification performance, Fig. 4.6b, d, f display the 3D visuals at an azimuth of $90°$ and a top view of $90°$. As shown in Fig. 4.6f, the data from faults 5, 13, 17, and 18 are intertwined, with nearly all data from fault 5 being obscured. In Fig. 4.6b2, the data from faults 5, 13, and 17 are mixed, while fault 18 is distinctly separated, showing clearer results compared to LPP. Figure 4.6b indicates that only a small portion of the data from faults 5 and 13 overlaps. Clearly, the datasets projected by the FDGLPP algorithm are more separated with less overlap than those from GLPP and LPP, demonstrating the distinct advantages of FDGLPP in terms of discrimination capability.

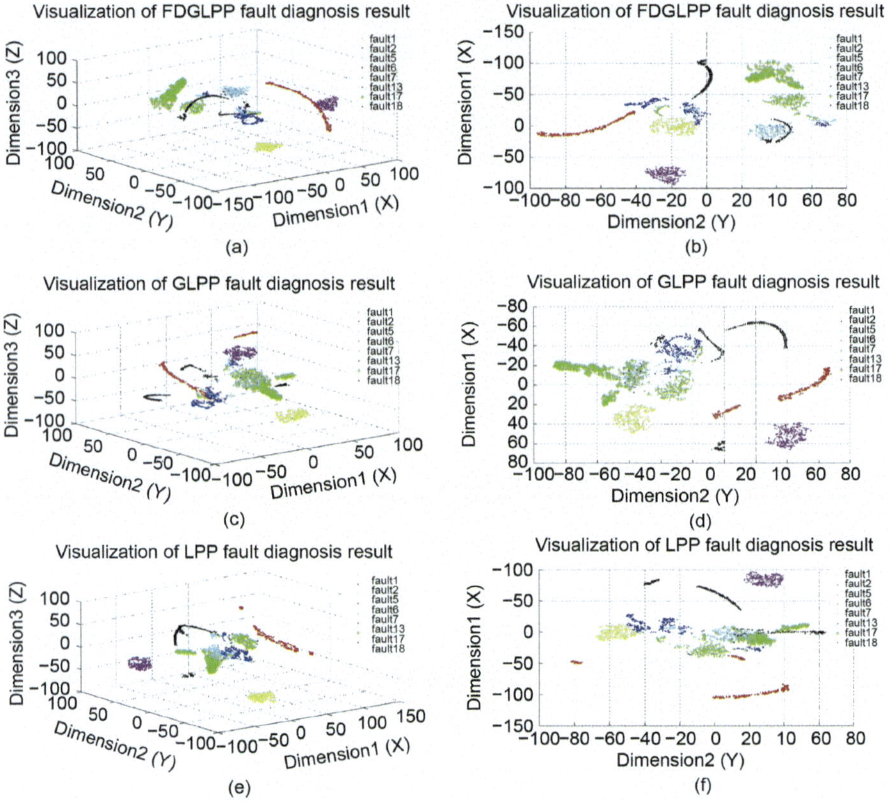

Fig. 4.6 Visualization of FDGLPP, GLPP and LPP classification results: (a1) FDGLPP, (b1) GLPP, (c1) LPP, and (a2), (b2), (c2) are the 3D view of figure a1, b1, c1 at an azimuth of 90 degrees and a top view of 90 degrees respectively

4.5 Conclusion

An innovative process monitoring method called FDGLPP is introduced, specifically tailored for fault detection and identification in industrial processes. This approach synergizes the benefits of Fisher discriminant analysis with global-local preserving projection, enabling it to preserve the overall Euclidean structure of the data while capturing local neighborhood characteristics. As a result, the transformation matrix generated possesses strong discriminative capabilities, enhancing fault identification performance. Moreover, the FDGLPP algorithm considers both manifold structure and discriminability, effectively tackling the challenges posed by overlapping fault data in multi-classification scenarios. To reduce the impact of noise and measurement errors encountered in practical processes, kernel density estimation is utilized to determine the control limits for monitoring. Consequently, FDGLPP extracts more information within the low-dimensional latent variable space, increasing its applicability in real-world situations. Numerical simulations and case studies indicate that this method leads to shorter detection delays and improved accuracy in fault diagnosis, confirming the superior performance and sensitivity of the FDGLPP algorithm in fault identification.

References

1. Bakshi BR (1998) Multiscale pca with application to multivariate statistical process monitoring. AIChE J 44(7):1596–1610
2. Wanke Yu, Zhao C (2017) Sparse exponential discriminant analysis and its application to fault diagnosis. IEEE Trans Indus Electron 65(7):5931–5940
3. Yang J, Yang J (2003) Why can lda be performed in pca transformed space? Pattern Recogn 36(2):563–566
4. Luo L, Bao S, Mao J, Ding Z (2018) Industrial process monitoring based on knowledge-data integrated sparse model and two-level deviation magnitude plots. Indus Eng Chem Res 57(2):611–622
5. Luo L (2014) Process monitoring with global-local preserving projections. Indus Eng Chem Res 53(18):7696–7705
6. Luo L, Bao S, Mao J, Tang D (2016) Nonlinear process monitoring based on kernel global-local preserving projections. J Process Control 38:11–21
7. Zhan C, Li S, Yang Y (2017) Enhanced fault detection based on ensemble global-local preserving projections with quantitative global-local structure analysis. Indus Eng Chem Res 56(38):10743–10755
8. Bao S, Luo L, Mao J, Tang D (2016) Improved fault detection and diagnosis using sparse global-local preserving projections. J Process Control 47:121–135
9. Chiang LH, Russell EL, Braatz RD (2000) Fault diagnosis in chemical processes using fisher discriminant analysis, discriminant partial least squares, and principal component analysis. Chemometrics Intell Lab Syst 50(2):243–252
10. Jiang B, Zhu X, Huang D, Paulson JA, Braatz RD (2015) A combined canonical variate analysis and fisher discriminant analysis (cva–fda) approach for fault diagnosis. Comput Chem Eng 77:1–9
11. Tang Q, Chai Y, Jianfeng Q, Ren H (2018) Fisher discriminative sparse representation based on dbn for fault diagnosis of complex system. Appl Sci 8(5):795
12. Zhang M, Ge Z, Song Z, Ruowei F (2011) Global-local structure analysis model and its application for fault detection and identification. Indus Eng Chem Res 50(11):6837–6848
13. Mugdadi AR, Ahmad IA (2004) A bandwidth selection for kernel density estimation of functions of random variables. Comput Stat Data Anal 47(1):49–62
14. Zhang K, Hao H, Chen Z, Ding SX, Peng K (2015) A comparison and evaluation of key performance indicator-based multivariate statistics process monitoring approaches. J Process Control 33:112–126
15. Van der Maaten L, Hinton G (2008) Visualizing data using t-sne. J Mach Learn Res 9(11)

Chapter 5
Dynamic Process Monitoring Based on Canonical Global and Local Preserving Projection Analysis

A novel framework that combines canonical variate analysis (CVA) with Global-Local Preserving Projection (GLPP) is proposed for fault monitoring in dynamic processes, referred to as canonical GLPP analysis. This method constructs a Laplacian matrix using a Hankel matrix, effectively capturing temporal correlations and underlying patterns. To solve the problem, Cholesky decomposition is applied to the covariance and cross-covariance of the Laplacian matrix, transforming it into a generalized eigenvalue problem. The resulting canonical GLPP analysis identifies an optimal projection matrix, uncovering dynamic variations by integrating both self-correlations and cross-correlations, while simultaneously preserving global and local structures as dictated by manifold learning. The performance of this method in fault detection and classification is demonstrated using a case study on the Tennessee Eastman process. Case study shows that the canonical GLPP analysis provides a more reliable and accurate framework for fault detection in dynamic processes.

5.1 Motivation

Driven by dynamics, uncontrollable disturbance and feedback loops in system, measurements from large scale continuous process involve both self-correlation and cross correlations. Thus, fault detection can be rather challenging for dynamic industrial processes. Canonical correlation analysis (CVA) has show powerful effectiveness on learning dynamic relations in dynamic processes [1]. CVA, however, is more suited for processes with self-correlation. For multivariate time series with significant cross-correlation, it becomes less effective. Instead, Global-Local Preserving Projection (GLPP) [2] not only preserves these structures but also enhances the discriminative power of the projection subspace, making it highly beneficial for fault detection. To this end, this chapter proposes a hybrid framework integrating CVA with GLPP. It merges time series analysis with spatial manifold structures to track dynamic processes. The benefits of this combination are in the following aspects:

H. Yin et al., *Data-Driven Fault Diagnosis for Complex Industrial Processes*, Engineering Applications of Computational Methods 22,
https://doi.org/10.1007/978-981-96-3153-7_5

- Maintaining both spatial structure and temporal information during dimension reduction. The canonical GLPP analysis method formulates the objective function using a Hankel matrix, which effectively captures the majority of dynamic variations within the data samples. The Laplacian matrix derived from this Hankel matrix, viewed through the lens of manifold structure, enables the optimal transformation matrix to leverage the strengths of CVA in managing high-dimensional data with serial correlations, as well as the benefits of the GLPP method in pattern classification.
- Broadening the scope of CVA from fault detection to fault diagnosis. Traditional CVA is primarily used to derive canonical variables or state-space models, subsequently employing statistical indices for fault detection. However, its lack of consideration for label information linked to datasets limits its use in fault classification. this chapter combines CVA with GLPP, utilizing Cholesky decomposition and singular value decomposition (SVD) to identify the optimal transformation matrix. The differences observed in the projection space can then be utilized to diagnose the types of faults.
- Targeting the dynamic challenges in process monitoring through projection preservation. The performance of the proposed method is evaluated against CVA-FDA, DPCA, GLPP, and dynamic GLPP. Key indicators for comparison include fault detection delay, the frequency of false alarms, and the accuracy of fault classification.

By combining the strengths of GLPP for multi-type fault classification and CVA for addressing dynamic characteristics in processes, this hybrid approach offers a more robust and accurate solution for both process data classification and monitoring.

5.2 Preliminary

5.2.1 Canonical Variate Analysis

As a data-driven methodology, canonical variate analysis aims to find out the transformation matrix through maximizes the correlation between past and future observation in process monitoring as shown in Fig. 5.1, then applied the transformation matrix to past observation to acquire canonical variables which preserve the serial correlations of the measurement.

Generally, the practical industrial process are nonlinear system, while such nonlinear character results it is difficult to deal with in process monitoring. However, at a stable normal operating point, the system can be approximately recognized as a linear system as follows:

$$\begin{cases} \mathbf{x}_{(t+1)} = \mathbf{A}\mathbf{x}_{(t)} + \mathbf{B}\mathbf{u}_{(t)} + \mathbf{w}_{(t)} \\ \mathbf{y}_{(t)} = \mathbf{C}\mathbf{x}_{(t)} + \mathbf{D}\mathbf{u}_{(t)} + \mathbf{v}_{(t)} \end{cases} \tag{5.1}$$

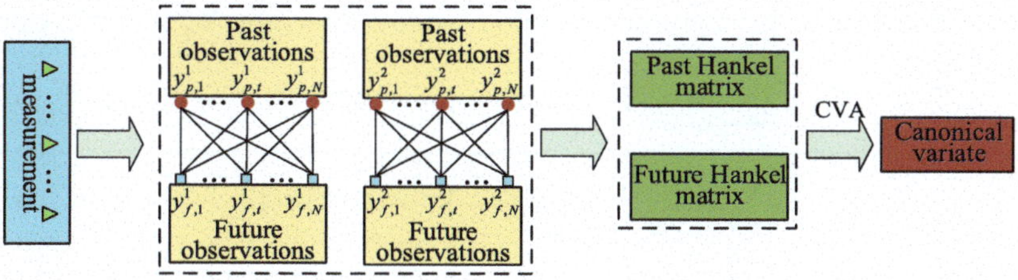

Fig. 5.1 Schematic of Canonical variate analysis

where the state variables $x_{(t)} \in R^{n_x}$ and output $\mathbf{y}_{(t)} \in R^{n_y}$ are acquired from the above multivariable stochastic state space model at sample point t, $\mathbf{A}$, $\mathbf{B}$, $\mathbf{C}$, $\mathbf{D}$ are system matrix, $\mathbf{w}_{(t)}$ is process disturb and $\mathbf{v}_{(t)}$ is measurement noise.

In the CVA method, the past observation matrix $\mathbf{y}_{p,t}$ and the future observation matrix $\mathbf{y}_{f,t}$ are constructed using past and future measurements of $\mathbf{y}_{(t)}$, incorporating lags of p and f, respectively. The appropriate lengths for these lags p and f can be identified by analyzing the self-correlation of the process variables [3].

$$\mathbf{y}_{p,t} = \begin{bmatrix} \mathbf{y}_{(t-1)} \\ \mathbf{y}_{(t-2)} \\ \vdots \\ \mathbf{y}_{(t-p)} \end{bmatrix} \in R^{pn_y}, \mathbf{y}_{f,t} = \begin{bmatrix} \mathbf{y}_{(t)} \\ \mathbf{y}_{(t+1)} \\ \vdots \\ \mathbf{y}_{(t+f-1)} \end{bmatrix} \in R^{fn_y} \tag{5.2}$$

In which t is a generic index of sample point. Assumes $t = m+1,\ m+2,\ \cdots m+N$, past and future Hankel matrix $\mathbf{Y}_p$ and $\mathbf{Y}_f$ is constructed as:

$$\mathbf{Y}_p = \left[\mathbf{y}_{p,m+1}\ \mathbf{y}_{p,m+2}\ \cdots\ \mathbf{y}_{p,m+N}\right] \in R^{pn_y \times N} \tag{5.3}$$

$$\mathbf{Y}_f = \left[\mathbf{y}_{f,m+1}\ \mathbf{y}_{f,m+2}\ \cdots\ \mathbf{y}_{f,m+N}\right] \in R^{fn_y \times N} \tag{5.4}$$

The subscript index of the last element in the first column of $\mathbf{Y}_p$ is $m+1-p$, the subscript index of the last element in the last column of $\mathbf{Y}_f$ is $m+N+f-1$. Since Hankel matrix is symmetric and constant across the anti-diagonals, given a data set with M samples of n_y measurement variables, M should satisfy the equation $M = (m+N+f-1) - (m+1-p) + 1$, thus the dimension of $\mathbf{Y}_p$ and $\mathbf{Y}_f$ is $N = M - f - p + 1$.

The estimation of the covariance and cross-covariance of $\mathbf{Y}_p$ and $\mathbf{Y}_f$ can be formulated as:

$$\sum\nolimits_{pp} = \mathbf{Y}_p\mathbf{Y}_p^T/(N-1) \in R^{mp \times mp} \tag{5.5}$$

$$\sum\nolimits_{ff} = \mathbf{Y}_f\mathbf{Y}_f^T/(N-1) \in R^{mf \times mf} \tag{5.6}$$

$$\sum\nolimits_{fp} = \mathbf{Y}_f\mathbf{Y}_p^T/(N-1) \in R^{mf \times mp} \tag{5.7}$$

From the perspective of dimensionality reduction, CVA seeks to identify the optimal linear combinations of past and future observations, denoted as $\mathbf{b}^T\mathbf{y}_{p,t}$ and $\mathbf{a}^T\mathbf{y}_{f,t}$, respectively. The goal is to maximize the correlation between these two sets of observations. In this context, $\mathbf{a} \in R^{fn_y}$ and $\mathbf{b} \in R^{pn_y}$ represent the transformation vectors applied to the future and past observations, respectively. The relationship between $\mathbf{a}^T\mathbf{y}_{f,t}$ and $\mathbf{b}^T\mathbf{y}_{p,t}$ is typically measured using the Pearson correlation coefficient $\rho_{fp}(\mathbf{a}, \mathbf{b})$:

$$\rho_{fp}(\mathbf{a}, \mathbf{b}) = \frac{\operatorname{cov}\left(\mathbf{a}^T\mathbf{y}_{f,t}, \mathbf{b}^T\mathbf{y}_{p,t}\right)}{\sigma_{\mathbf{a}^T\mathbf{y}_{f,t}}\sigma_{\mathbf{b}^T\mathbf{y}_{p,t}}} = \frac{\mathbf{a}^T\sum_{fp}\mathbf{b}}{\left(\mathbf{a}^T\sum_{ff}\mathbf{a}\right)^{\frac{1}{2}}\left(\mathbf{b}^T\sum_{pp}\mathbf{b}\right)^{\frac{1}{2}}} \tag{5.8}$$

In which $\sigma_{\mathbf{a}^T\mathbf{y}_{f,t}}$ is the variance of $\mathbf{y}_{f,t}$, $\sigma_{\mathbf{b}^T\mathbf{y}_{p,t}}$ is the variance of $\mathbf{y}_{p,t}$. By define $\mathbf{u} = \sum_{ff}^{-\frac{1}{2}}\mathbf{a}$, $\mathbf{v} = \sum_{ff}^{-\frac{1}{2}}\mathbf{b}$, the optimization to calculated $\mathbf{a}$ and $\mathbf{b}$ can be formulated as:

$$\begin{gathered}\max_{\mathbf{u},\mathbf{v}} \mathbf{u}^T\left(\sum\nolimits_{ff}^{-\frac{1}{2}}\sum\nolimits_{fp}\sum\nolimits_{pp}^{-\frac{1}{2}}\right)\mathbf{v} \\ s.t.\mathbf{u}^\mathbf{T}\mathbf{u} = 1, \mathbf{v}^T\mathbf{v} = 1\end{gathered} \tag{5.9}$$

A limitation in this formulation lies in scaling the denominator in Eqs. (5.1) to (5.9). This optimization problem can be addressed by applying SVD to the Hankel matrix $\mathbf{H}$:

$$\mathbf{H} := \sum_{ff}^{-\frac{1}{2}}\sum\nolimits_{fp}\sum_{pp}^{-\frac{1}{2}} = \mathbf{U}\mathbf{D}\mathbf{V}^T \in R^{mf \times mp} \tag{5.10}$$

$$\begin{aligned}\mathbf{U} &= \left[\mathbf{u}_1\ \mathbf{u}_2\ \cdots\ \mathbf{u}_r\right] \in R^{mf \times r} \\ \mathbf{V} &= \left[\mathbf{v}_1\ \mathbf{v}_2\ \cdots\ \mathbf{v}_r\right] \in R^{mp \times r}\end{aligned} \tag{5.11}$$

In this context, r represents the dimension of the Hankel matrix $\mathbf{H}$, $\mathbf{D}$ is a diagonal matrix consisting of the r singular values. The elements of $\mathbf{D}$ are known as the canonical correlation coefficients. The matrices $\mathbf{U}$ and $\mathbf{V}$ are orthogonal, containing the left singular vectors $\mathbf{u}_i, i = 1, 2, \ldots, r$ and the right singular vectors $\mathbf{v}_i, i = 1, 2, \ldots, r$, respectively

The two corresponding sets of canonical variables $\mathbf{c}_t$ and $\mathbf{d}_t$ can be formulated as:

$$\begin{aligned} \mathbf{c}_t &= \begin{bmatrix} \mathbf{b}_1^T \\ \mathbf{b}_2^T \\ \vdots \\ \mathbf{b}_r^T \end{bmatrix} \mathbf{y}_{p,t} = \mathbf{V}^T \sum\nolimits_{pp}^{-\frac{1}{2}} \mathbf{y}_{p,t} = \mathbf{J}\mathbf{y}_{p,t} \in R^r \\ \mathbf{d}_t &= \begin{bmatrix} \mathbf{a}_1^T \\ \mathbf{a}_2^T \\ \vdots \\ \mathbf{a}_r^T \end{bmatrix} \mathbf{y}_{f,t} = \mathbf{U}^T \sum\nolimits_{ff}^{-\frac{1}{2}} \mathbf{y}_{f,t} = \mathbf{L}\mathbf{y}_{f,t} \in R^r \end{aligned} \tag{5.12}$$

In which $\mathbf{J} = \mathbf{V}^T \sum_{pp}^{-\frac{1}{2}}$ and $\mathbf{L} = \mathbf{U}^T \sum_{ff}^{-\frac{1}{2}}$ are transformation matrix for past and future observations, respectively, they also named canonical loading matrix, $\sum_{pp}^{-\frac{1}{2}}$ and $\sum_{ff}^{-\frac{1}{2}}$ are introduced to guarantee canonical variables with unit variance. The covariance for the canonical variable $\mathbf{c}_t$ and $\mathbf{d}_t$, the cross-covariance between them are $\sum_{cc} = \mathbf{I}$, $\sum_{dd} = \mathbf{I}$, $\sum_{cd} = \mathbf{D}$, respectively. This shows the canonical variables $\mathbf{c}_t$ is most correlated with the corresponding canonical variables $\mathbf{d}_t$, and the optimal estimation for $\mathbf{d}_t$ is $\mathbf{D}\mathbf{c}_t = \mathbf{D}\mathbf{J}\mathbf{y}_{p,t}$, then the estimation residual $\delta_t = \mathbf{d}_t - \mathbf{D}\mathbf{c}_t = \mathbf{L}\mathbf{y}_{f,t} - \mathbf{D}\mathbf{J}\mathbf{y}_{p,t}$.

The canonical variables are arranged based on their predictive capability, with the first k dimensions offering the optimal representation of the system states achieved through the maximization of the likelihood function [4]. Following this, the estimation of the state variables will be performed.$\widehat{\mathbf{x}}_{(t)}$ at t can be represented as:

$$\widehat{\mathbf{x}}_{(t)} = \mathbf{J}_k \mathbf{y}_{p,t} = \begin{bmatrix} \mathbf{I}_k & \mathbf{0} \end{bmatrix} \mathbf{J}\mathbf{y}_{p,t} \tag{5.13}$$

$$\mathbf{r}_t = \left(I - \mathbf{V}_x \mathbf{V}_x^T\right) \sum_{pp}^{-\frac{1}{2}} \mathbf{y}_{p,t} \tag{5.14}$$

In this context, the notation is referring to various components involved in the formulation of a state estimation or a dimensionality reduction problem. Here's a clearer breakdown and translation of the text: $\mathbf{J}_k$ consists of the first k columns of the transformation matrix $\mathbf{J}$. $\mathbf{V}_x$ consists of the first k columns of the matrix $\mathbf{V}$. $\mathbf{r}_t$ is the residual space, which typically refers to the part of the data or signal that is left unexplained by the projection onto the lower-dimensional space. $\widehat{\mathbf{x}}_{(t)}$ is the estimated state of the system at time t, obtained through the transformation matrix $\mathbf{J}$ and past observations. It is evident that the state space is a subset of the canonical space, and is also a linear combination of past observations.

5.2.2 Global-Local Preserving Projection

The GLPP algorithm is designed to capture spatial correlations among variables by preserving both the local neighborhood structure and the global manifold structure [5], as illustrated in Fig. 5.2. This figure displays a one-dimensional projection of a normal dataset **X** in two dimensions, with the blue, yellow, and green intervals representing projections along the GPP (global-preserving projection), LPP (local preserving projection), and GLPP directions, respectively. The yellow points x_1, x_2 and x_3 indicate fault data. It is evident that the LPP direction only detects x_1, while the GPP direction can identify only x_2. In contrast, the GLPP direction successfully detects all three fault data points: x_1, x_2 and x_3. Denote the adjacent and non-adjacent relationships between $\mathbf{x}_i$ and $\mathbf{x}_j$ from data set $\mathbf{X} \in \mathbb{R}^{m\times n}$ is w_{ij} and $\bar{w}_{ij}$, respectively. w_{ij} constructs the local neighborhood weight matrix **W**, and $\bar{w}_{ij}$ constructs the global weight matrix $\bar{W}$. GLPP aims to find out a set of vectors $\mathbf{a}_i \in \mathbb{R}^{d\times 1}$, which consist transformation matrix $\mathbf{A} \in \mathbb{R}^{m\times d}$, $d < m$. This matrix can project original data set into dimension reduced space. w_{ij} and $\bar{w}_{ij}$ is defined as:

$$w_{ij} = \begin{cases} e^{-\frac{\|\mathbf{x}_i - \mathbf{x}_j\|^2}{\sigma_1}}, if\ \mathbf{x}_j \in \Omega_k(\mathbf{x}_i) \text{ or } \mathbf{x}_i \in \Omega_k(\mathbf{x}_j) \\ 0\text{ , otherwise} \end{cases} \tag{5.15}$$

$$\bar{w}_{ij} = \begin{cases} e^{-\frac{\|\mathbf{x}_i - \mathbf{x}_j\|^2}{\sigma_2}}, if\ \mathbf{x}_j \notin \Omega_k(\mathbf{x}_i) \text{ and } \mathbf{x}_i \notin \Omega_k(\mathbf{x}_j) \\ 0\text{ , otherwise} \end{cases} \tag{5.16}$$

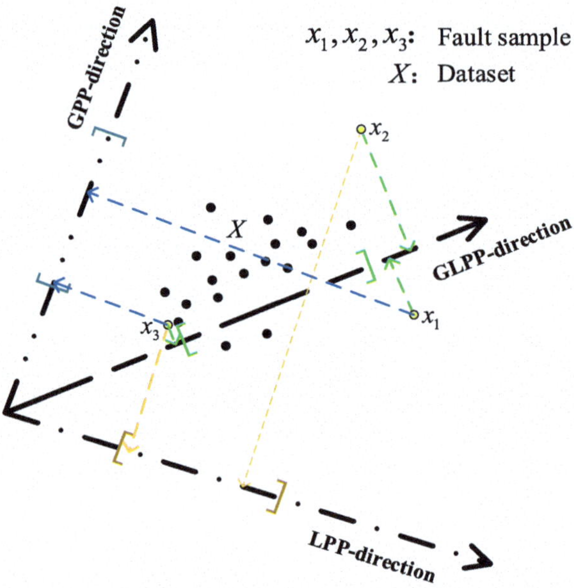

Fig. 5.2 Visualization of LPP, GPP and GLPP method performance on 2-D dataset with 1-D direction

where σ_1 and σ_2 is constant, and $\Omega_k(\mathbf{x}_i)$ is the collection of k nearest neighbors of data sample $\mathbf{x}_i$.

The objective function can be defined as:

$$J_{GLPP}(\mathbf{a}) = \min_{\mathbf{a}} \frac{1}{2}\left\{\sum\nolimits_{i,j}\begin{pmatrix}\eta\left(y_i - y_j\right)^2 W_{ij} \\ -(1-\eta)\sum\nolimits_{i,j}\left(y_i - y_j\right)^2 \bar{W}_{ij}\end{pmatrix}\right\} \tag{5.17}$$

where scalar y_i is the projection of vector $\mathbf{x}_i$ by $y_i = \mathbf{a}^T\mathbf{x}_i$, $\mathbf{a}$ is the transformation vector in $\mathbf{A}$, Eq. (5.17) can be rewritten as:

$$\begin{aligned} J_{GLPP}(\mathbf{a}) &= \min_{\mathbf{a}} \tfrac{1}{2}\left\{\sum\nolimits_{i,j}\begin{pmatrix}\eta\left(y_i - y_j\right)^2 W_{ij} \\ -(1-\eta)\sum\nolimits_{i,j}\left(y_i - y_j\right)^2 \bar{W}_{ij}\end{pmatrix}\right\} \\ &= \min_{\mathbf{a}} \tfrac{1}{2}\left\{\sum\nolimits_{i,j}\left(\left(y_i - y_j\right)^2 R_{ij}\right)\right\} \\ &= \min_{\mathbf{a}}\left\{\sum\nolimits_{i}\left(\mathbf{a}^T\mathbf{x_i}\mathbf{H_{ii}}\mathbf{x}_i^T\mathbf{a}\right) - \sum\nolimits_{i,j}\left(\mathbf{a}^T\mathbf{x}_i\mathbf{R}_{ij}\mathbf{x}_j^T\mathbf{a}\right)\right\} \\ &= \min_{\mathbf{a}}\left\{a^T\mathbf{X}(\mathbf{H}-\mathbf{R})\mathbf{X}^T a\right\} \\ &= \min_{\mathbf{a}}\left\{a^T\mathbf{X}\mathbf{M}\mathbf{X}^T a\right\} \end{aligned} \tag{5.18}$$

In this formulation, $Rij = \eta Wij - (1-\eta)\bar{W}ij$, where $\mathbf{H}$ is a diagonal matrix with diagonal entries $Hii = \sum_j R_{ij}$. The Laplacian matrix is defined as $\mathbf{M} = \mathbf{H} - \mathbf{W}$. The constant η is introduced to control the trade-off between preserving local and global information during the projection.

Furthermore, to normalize the solution, a constraints is introduced to above model, then the objective function of GLPP can be described as [2]:

$$J_{GLPP}(\mathbf{a}) = \min_{\mathbf{a}}\left\{\mathbf{a}^{\mathbf{T}}\mathbf{X}\mathbf{M}\mathbf{X}^{\mathbf{T}}\mathbf{a}\right\} \text{ s.t. } \mathbf{a}^{\mathbf{T}}\mathbf{X}\mathbf{N}\mathbf{X}^{\mathbf{T}}\mathbf{a} = 1 \tag{5.19}$$

where $\mathbf{N} = \eta\mathbf{X}\mathbf{H}\mathbf{X}^T + (1-\eta)\mathbf{I}_m$.

The above optimization problem can be transformed to a generalized eigenvalue problem as:

$$\mathbf{X}\mathbf{M}\mathbf{X}^T\mathbf{a} = \lambda\mathbf{X}\mathbf{N}\mathbf{X}^T\mathbf{a} \tag{5.20}$$

The optimal GLPP transformation matrix $\mathbf{A}$ is consist of eigenvectors $a_1, a_2, \ldots, a_l$, the sequence is sorted by the corresponding eigenvalues $\lambda_1 < \lambda_2 <, \ldots, < \lambda_l$.

5.3 Methodology

5.3.1 *Global-Local Preserving Projection Based on Canonical Variate Analysis*

As industrial processes grow in scale and complexity, the manufacturing methods and kinetic mechanisms become increasingly intricate. This results in a higher number of variables and samples, along with complicated interconnections involving both spatial and temporal correlations. While CVA effectively captures the self-correlation among monitoring variables, it primarily focuses on the temporal characteristics of the data, neglecting the spatial structure. Furthermore, most research on CVA in the context of process monitoring has predominantly centered on fault detection, with fewer studies addressing fault diagnosis through CVA methodologies.

A hybrid framework combining CVA with GLPP is proposed to create a robust and accurate process data monitoring and fault detection system. Instead of constructing the objective function using the traditional monitoring matrix, this approach employs a Hankel matrix, which captures self-correlation information, as shown in Fig. 5.3. The global and local preserving analysis is applied to generate past and future Laplacian matrices from the Hankel matrix. Cholesky decomposition is then used to reformulate the covariance and cross-covariance of these matrices, converting the problem into one of singular value decomposition. This method produces a transformation matrix for canonical variables that effectively maintains both spatial and temporal relationships within the data. Additionally, by integrating label information during the projection, the approach significantly improves fault detection and classification accuracy.

Apply GLPP model in 5.2.2. to past Hankel matrix $\mathbf{Y}_p$ consist of the past observations $\mathbf{y}_{p,t}$, then the objective function of GLPP based on CVA can be described as:

$$J_{GLPP}(\mathbf{a}) = \min_{\mathbf{a}} \frac{1}{2} \left\{ \sum\nolimits_{i,j} \begin{pmatrix} \eta (y_i - y_j)^2 W_{ij} \\ -(1-\eta) \sum\nolimits_{i,j} (y_i - y_j)^2 \bar{W}_{ij} \end{pmatrix} \right\} \tag{5.21}$$

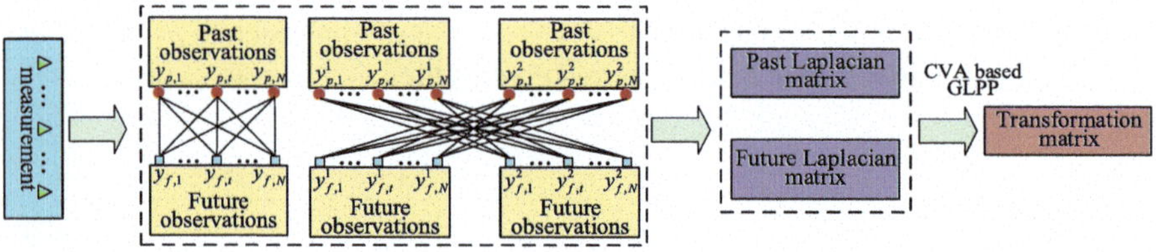

Fig. 5.3 Visualization of LPP, GPP and GLPP method performance on 2-D dataset with 1-D direction

In which scalar y_i is the projection of vector $\mathbf{y}_{p,t}$ by $y_i = \mathbf{w}_p^T \mathbf{y}_{p,t}$, $\mathbf{w}_p$ is the transformation vector for past observations in $\mathbf{W}_p$. W_{ij} and $\bar{W}_{ij}$ is defined as the adjacent and non-adjacent relationships between past observations, respectively.

According to Eqs. (5.18), (5.21) can be rewritten as:

$$\begin{aligned} J_{GLPP}\left(\mathbf{w}_p\right) &= \min_{\mathbf{a}} \tfrac{1}{2} \left\{ \sum\nolimits_{i,j} \begin{pmatrix} \eta\left(y_i - y_j\right)^2 W_{ij} \\ -(1-\eta)\sum\nolimits_{i,j}\left(y_i - y_j\right)^2 \bar{W}_{ij} \end{pmatrix} \right\} \\ &= \min_{\mathbf{a}} \left\{ \mathbf{w}_p^T \mathbf{Y}_p \left(\mathbf{H} - \mathbf{R}\right) \mathbf{Y}_P^T \mathbf{w}_p \right\} \\ &= \min_{\mathbf{a}} \left\{ \mathbf{w}_p^T \mathbf{Y}_p \mathbf{M}_p \mathbf{Y}_P^T \mathbf{w}_p \right\} \end{aligned} \tag{5.22}$$

In which $\mathbf{M}_p$ is the Laplacian matrix of the past observations, and the Laplacian matrix of the future observations $\mathbf{M}_f$ is defined analogously as:

$$J_{CGLPP}\left(\mathbf{w}_f\right) = \min_{\mathbf{a}} \left\{ \mathbf{w}_f^T \mathbf{Y}_f \mathbf{M}_f \mathbf{Y}_f^T \mathbf{w}_f \right\} \tag{5.23}$$

In which $\mathbf{w}_f$ is the transformation vector for future observations in $\mathbf{W}_f$.

Since $\mathbf{M}_p$ is semi-definite, applying Cholesky decomposition on $\mathbf{M}_p$ and defined $\mathbf{C}_{pp}$ as:

$$\begin{aligned} \mathbf{C}_{pp} &= \mathbf{Y}_p \mathbf{M}_p \mathbf{Y}_p^T = \mathbf{Y}_p \left(\mathbf{N}_p^T \Lambda_p \mathbf{N}_p\right) \mathbf{Y}_p^T \\ &= \left(\mathbf{Y}_p \mathbf{N}_p^T \Lambda_p^{\frac{1}{2}}\right)\left(\Lambda_p^{\frac{1}{2}} \mathbf{N}_p \mathbf{Y}_p^T\right) = \left(\mathbf{Y}_p \mathbf{N}_p^T \Lambda_p^{\frac{1}{2}}\right)\left(\mathbf{Y}_p \mathbf{N}_p^T \Lambda_p^{\frac{1}{2}}\right)^T \end{aligned} \tag{5.24}$$

In which $\mathbf{N}_p$ is upper triangular matrix, Λ_p is diagonal matrix, $\mathbf{S}_p = \mathbf{N}_p^T \Lambda_p^{\frac{1}{2}}$. And $\mathbf{C}_{ff}$, $\mathbf{S}_f$ can be defined analogously. Obviously, $\mathbf{C}_{pp}$ and $\mathbf{C}_{ff}$ can be reformulated as the form with covariance.

Rather than estimating the covariance and cross-covariance of the past and future Hankel matrices, the canonical GLPP analysis focuses on estimating the covariance and cross-covariance of the Laplacian matrix. Based on Eq. (5.18), the objective of the canonical GLPP analysis algorithm can be expressed as follows:

$$\max_{\mathbf{w}_p, \mathbf{w}_f} \frac{\mathbf{w}_p^T \mathbf{C}_{fp} \mathbf{w}_f}{\sqrt{\mathbf{w}_p^T \mathbf{C}_{pp} \mathbf{w}_p} \cdot \sqrt{\mathbf{w}_f^T \mathbf{C}_{ff} \mathbf{w}_f}} \tag{5.25}$$

In which $\mathbf{C}_{fp} = \mathbf{Y}_f \mathbf{S}_f \mathbf{S}_p^T \mathbf{Y}_p^T$. This optimization aims to minimize local spatial correlations while simultaneously maximizing global correlations among observations from both past and future data. Additionally, it takes serial correlations into account to enhance the canonical correlations between past and future observations. Therefore, the objective function for the GLPP-based CVA algorithm can be expressed as:

$$\max_{\mathbf{w}_p, \mathbf{w}_f} \mathbf{w}_p^T \mathbf{C}_{fp} \mathbf{w}_f s.t. \mathbf{w}_p^T \mathbf{C}_{pp} \mathbf{w}_p = 1, \mathbf{w}_f^T \mathbf{C}_{ff} \mathbf{w}_f = 1 \tag{5.26}$$

Following the standard procedures of CVA in Sect. 5.2.1, solving the optimal transformation vector $\mathbf{w}_p$ and $\mathbf{w}_f$ can be determined as singular value decomposition of the Hankel matrix defined by $\mathbf{H} := \mathbf{C}_{ff}^{-\frac{1}{2}}\mathbf{C}_{fp}\mathbf{C}_{pp}^{-\frac{1}{2}} = \mathbf{UDV}^T$. Canonical variables and residual variables based on the optimal transformation matrix can be solved by Eq. (5.18) and determined as:

$$\begin{aligned}
\mathbf{Z}_t &= \mathbf{W}_p\mathbf{Y}_p\mathbf{S}_p = \begin{bmatrix} \mathbf{w}_{p,1}^T \\ \mathbf{w}_{p,2}^T \\ \vdots \\ \mathbf{w}_{p,r}^T \end{bmatrix} \mathbf{Y}_p\mathbf{S}_p = \mathbf{V}^T\mathbf{C}_{pp}^{-\frac{1}{2}}\mathbf{Y}_p\mathbf{S}_p \\
\mathbf{D}_t &= \mathbf{W}_f\mathbf{Y}_f\mathbf{S}_f = \begin{bmatrix} \mathbf{w}_{f,1}^T \\ \mathbf{w}_{f,2}^T \\ \vdots \\ \mathbf{w}_{f,r}^T \end{bmatrix} \mathbf{Y}_f\mathbf{S}_f = \mathbf{U}^T\mathbf{C}_{ff}^{-\frac{1}{2}}\mathbf{Y}_f\mathbf{S}_f
\end{aligned} \tag{5.27}$$

In which $\mathbf{U}$ and $\mathbf{V}$ are orthogonal matrix consist of left singular matrix and right singular matrix of Hankel matrix, respectively.

Monitoring statistic index T_c^2 from canonical variables $\mathbf{c}_t$ and Q_c from residual variables $\mathbf{d}_t$ are introduced to detect whether fault occurs in process. $\mathbf{d}_t$ is the vector of $\mathbf{D}_t$, The T_c^2 and Q_c is defined as:

$$T_c^2 = \mathbf{c}_t^T\mathbf{S}^{-1}\mathbf{c}_t \tag{5.28}$$

$$Q_c = \mathbf{d}_t^T\mathbf{d}_t \tag{5.29}$$

In this expression, $S = \mathbf{C}_t^T\mathbf{C}_t/(n-1)$, where $\mathbf{c}_t$ represents the columns of $\mathbf{C}_t$. The statistic T_c^2 quantifies the magnitude of variations within the canonical space, making it effective for detecting faults with significant magnitudes. In contrast, Q_c captures variations in the residual space, allowing it to be more sensitive to faults with smaller magnitudes.

5.3.2 Control Limit Through Kernel Density Estimation

To establish the control limits for the T_c^2 and Q_c statistics, traditional approaches typically assume that T_c^2 follows a weighted F-distribution, while Q_c follows a weighted χ^2-distribution. However, due to the dynamic nature of real-world processes, these assumptions may not hold, leading to decreased accuracy and sensitivity in process monitoring. To address this issue, this chapter introduces KDE as a method for determining the control limits for the proposed statistical indices, accounting for the dynamics observed in practical processes.

The T_c^2 of all training samples $\mathbf{X} = [\mathbf{x}_1, \mathbf{x}_2, \ldots \mathbf{x}_n]$ is defined as $\left[T_1^2, T_2^2, \ldots T_n^2\right]$, then the probability density function of T_c^2 is :

$$\hat{f}(t) = \frac{1}{nh} \sum_{i=1}^{n} K\left(\frac{t - T_i^2}{h}\right) \tag{5.30}$$

where n represents the number of data samples, and the kernel function $K(\cdot)$ defines the shape of the bumps, with a Gaussian kernel function chosen for this chapter. The bandwidth h determines the width of the bumps, which can be determined using the least-squares cross-validation method, as given by $h = 1.06\sigma n^{-0.2}$ [6, 7], where σ denotes the standard deviation of the sample.

The cumulative density function $P(x < \theta)$ is obtained by integrating probability density function $\hat{f}(t)$ as follows:

$$P(x < \theta) = \int_{-\infty}^{\theta} \hat{f}(x)dx = 1 - \alpha \tag{5.31}$$

Choosing the confidence level $\alpha = 0.01$, then the control limit of T_c^2 can be determined by calculated the upper limit of the integration in Eq. (5.31). The control limit of Q_c index is computed in the same way as T_c^2 statistics. Once the T_c^2 or Q_c statistics of a new sample exceeds the control limit, it is defined that fault occurs in system.

5.3.3 *Parameter Selection*

5.3.3.1 Time Lags p and f

The time lags p and f play a vital role in defining the canonical variate space. While increasing these lags can enhance the accuracy of the models produced by the proposed method, it may also lead to overfitting and reduce the robustness of the monitoring techniques. Moreover, greater lag values necessitate more data, resulting in higher computational costs for parameter estimation [8]. Conversely, using excessively small lags may fail to capture the necessary serial correlations. Variance and covariance only describe the probabilistic properties of a stochastic process at individual time points and do not adequately represent the underlying relationships within the process. To address this limitation, the self-correlation function—a specific instance of cross-correlation—is employed to determine appropriate values for p and f. The self-correlation quantifies the relationship between a signal and its delayed version across various lags, using the Pearson correlation coefficient, and it is defined as follows:

$$\rho_{\mathbf{yy}}(\tau) = \frac{E\left[\left(\mathbf{y} - \mu_{\mathbf{y}}\right)\left(\mathbf{y}_{\tau} - \mu_{\mathbf{y}_{\tau}}\right)\right]}{\sigma_{\mathbf{y}}\sigma_{\mathbf{y}_{\tau}}} \tag{5.32}$$

In which $\mathbf{y}_{\tau}$ is the delayed version of process measurement $\mathbf{y}$ with τ lags. The mean and variance of $\mathbf{y}$ and $\mathbf{y}_{\tau}$ are $\mu_{\mathbf{y}}$, $\mu_{\mathbf{y}_{\tau}}$, $\sigma_{\mathbf{y}}$, $\sigma_{y_{\tau}}$, respectively. According to this, it is rational to determine the time lags that offer significant correlation.

τ is determined by the minimum self-correlation of the squared sum of M variables, as shown in Eq. (5.33). For simplicity, assume that $\tau = h = l$. When the time distance exceeds this threshold, the correlation is considered negligible [3]. The self-correlation of the training dataset is evaluated with a confidence bound of $\pm 5\%$.

$$\mathbf{s}_i = \sum_{j=1}^{n_y} \mathbf{y}_{i,j}^2 \quad \min_{\tau} \rho_{\mathbf{s}_i\mathbf{s}_i}(\tau) \tag{5.33}$$

In which $\mathbf{y}_{i,j}$ is the j_{th} element of sample $\mathbf{y}_i$, $\rho_{\mathbf{y}_i\mathbf{y}_i}(\tau) = 0$, once $i + \tau > M$.

5.3.3.2 The Reduced Dimension *k*

The process of selecting the reduced dimension can be viewed as determining the state order of the canonical space, which serves as the foundation for monitoring processes using CVA. Three commonly used methods for this selection include: first, a technique based on the magnitude of singular values, where the top k dominant canonical variates are chosen as state variables according to a predefined threshold [9]. This method is particularly effective when the noise level in the process is relatively low compared to the monitoring variables, resulting in the first k singular values being significantly larger than the others. The second method involves cross-validation, where validation data is introduced to assess model performance. A third approach is to select the order that minimizes the AIC [10], a widely used metric for model selection. Given that noise in the process cannot be overlooked and the singular values decreased too gradually in the case study, this chapter utilizes the AIC to determine the appropriate order for the state space.

KLD measures the divergence between an approximation model and the true process. A smaller KLD indicates a better fit to the actual data. However, since the true process model is typically unknown, direct computation of the KLD is not possible. Drawing from the relationship between information loss and KLD [11], Akaike recognized a connection between the log likelihood function and KLD. This led to the formulation of the Akaike Information Criterion (AIC), a tool for statistical model selection based on observed data. The log likelihood function, expressed in terms of the residual δ, is defined as follows [12]:

$$-2\log p\left(\mathbf{Y}^M, \mathbf{U}^M \hat{\theta}_k\right) = -N\log\left|\mathbf{I} - \mathbf{D}^2\right| + \sum_{1}^{M} \delta_t^T \left(\mathbf{I} - \mathbf{D}^2\right)^{-1} \delta_t \tag{5.34}$$

In which the covariance of residual δ_t is $\sum_{\delta\delta} = \mathbf{I} - \mathbf{D}$. This equation shows the canonical variate estimation of the system parameters are approximately maximum likelihood estimation with optimal statistical properties.

The AIC for each order k is defined by:

$$AIC\,(k) = -2 \log p\left(\mathbf{Y}^M, \mathbf{U}^M \hat{\theta}_k\right) + 2q \tag{5.35}$$

$$q = k\,(2q_{out} + q_{in}) + q_{out}q_{in} + \frac{1}{2}q_{out}\,(q_{out} + 1) \tag{5.36}$$

In which $\left(\mathbf{Y}^M, \mathbf{U}^M\right)$ is the set of M observations, and $\mathbf{Y}^M = \left\{y_{(1)}, y_{(2)}, \ldots, y_{(M)}\right\}$, $\mathbf{U}^M = \left\{\mathbf{u}_{(1)}, \mathbf{u}_{(2)}, \ldots, \mathbf{u}_{(M)}\right\}$, $p\left(\mathbf{Y}^M, \mathbf{U}^M \hat{\theta}_k\right)$ is the likelihood function based on observations $\left(\mathbf{Y}^M, \mathbf{U}^M\right)$, $\hat{\theta}_k$ is the maximum likelihood parameter estimation using a k order model with q parameters, q_{in} and q_{out} are the number of inputs and outputs, respectively, k is determined according to the minimum of $AIC\,(k)$.

5.3.3.3 The Balance Parameter η of Global and Local Structure

To maintain an optimal balance between global and local structures in the proposed projection model, the spectral radius is employed as an indicator of structural information. The balance parameter of the GLPP model is then determined by applying the spectral radius to the local and global Laplacian matrices, which are derived from both past and future observations.

$$\eta = \frac{\rho\left(L_{pglobal}\right) + \rho\left(L_{fglobal}\right)}{\rho\left(L_{pglobal}\right) + \rho\left(L_{plocal}\right) + \rho\left(L_{fglobal}\right) + \rho\left(L_{flocal}\right)} \tag{5.37}$$

In which $\rho\,(.)$ is spectral radius of a matrix, $L_{pglobal}$ and L_{plocal} are the global and local Laplacian matrices of past observation matrix, respectively. $L_{fglobal}$ and L_{flocal} are the global and local matrices of future observation matrix, respectively.

5.3.4 *The Overall Framework*

As illustrated in Fig. 5.4, the off-line training and modeling process involves the following steps:

(1) Data collected during normal operational conditions is designated as training data. Z-score standardization is then applied to this training data to mitigate the effects of larger absolute values.

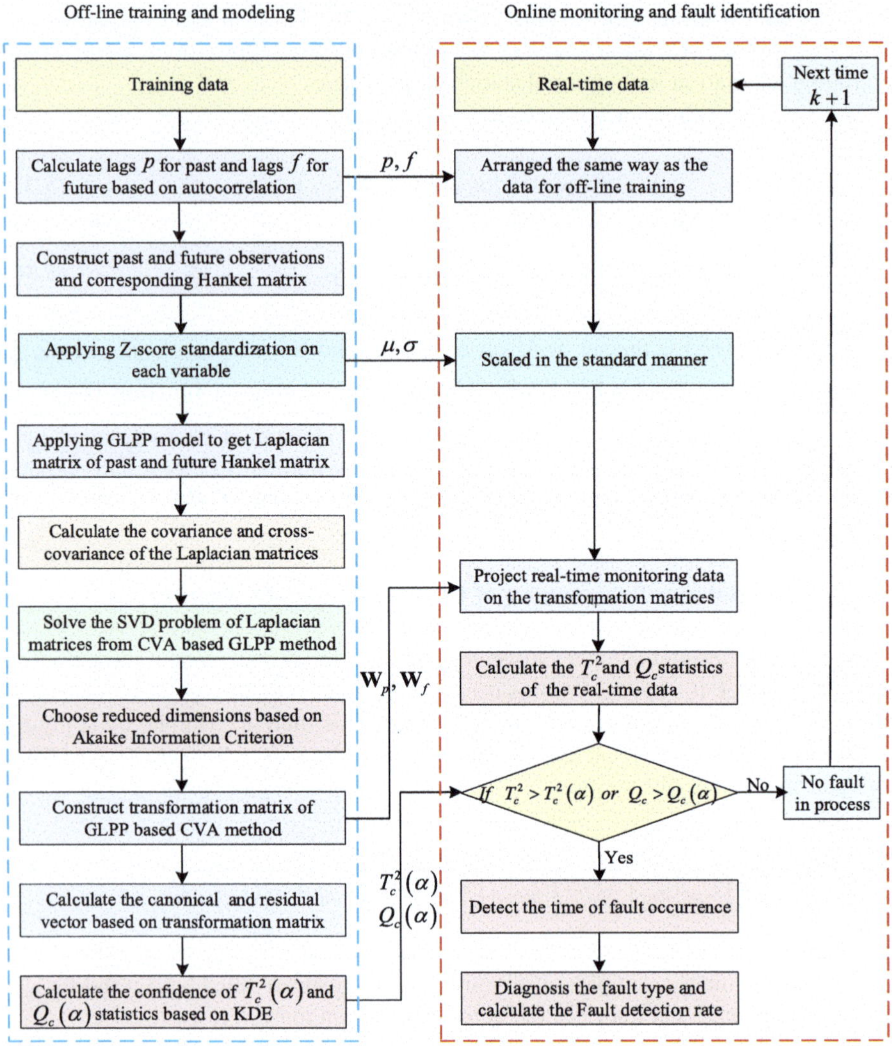

Fig. 5.4 Process monitoring procedure of Canonical GLPP analysis method

(2) Calculate lags p for past and lags f for future based on self-correlation of monitoring data according to Eqs. (5.33) and (5.34), then construct past observations $\mathbf{y}_{p,t}$ and future observations $\mathbf{y}_{f,t}$, respectively, according to Eq. (5.2).
(3) Construct past and future Hankel matrix $\mathbf{Y}_p$ and $\mathbf{Y}_f$ according to Eqs. (5.3) and (5.4), then applying GLPP model as Eq. (5.22) to get Laplacian matrices $\mathbf{M}_p$ and $\mathbf{M}_f$ of past and $\mathbf{M}_f$ future observations, respectively.
(4) Calculate the covariance matrices of Laplace matrix $\mathbf{M}_p$ and $\mathbf{M}_f$, and cross-covariance matrix between $\mathbf{M}_p$ and $\mathbf{M}_f$ as Eq. (5.24).

(5) Solve the SVD problem of Hankel matrix from GLPP based CVA model.
(6) Choose the reduced dimensions based on Akaike Information Criterion as Eq. (5.35), and construct transformation matrix $\mathbf{W}_p$ and $\mathbf{W}_f$ of GLPP based CVA method according to Eq. (5.27).
(7) Calculate control limit of index $T_c^2(\alpha)$ and $Q_c(\alpha)$ of canonical and residual vectors based on Kernel density estimation according to Eq. (5.31).

Online monitoring and fault identification:

(1) Collect real-time data and scale in the standard manner using the mean μ and variance σ of training data
(2) Arranged the same way as the data for off-line training to construct the past observation and future observation for current time.
(3) According to Eq. (5.27) Project real-time monitoring data on the transformation matrix $\mathbf{W}_p$ for past observations and $\mathbf{W}_f$ for future observations, respectively, which can both preserve the serial and spatial information simultaneously.
(4) Calculate T_c^2 and Q_c statistics of the real-time data according to Eqs. (5.28) and (5.29).
(5) Fault is detected if $T_c^2 > T_c^2(\alpha)$ or $Q_c > Q_c(\alpha)$, then detect the time of fault occurrence and diagnosis the fault type.
(6) If there are no fault then collect data of next time and return to step (1).

5.4 Case Study

5.4.1 Tennessee Eastman Process

The TEP is used to assess the effectiveness of the proposed canonical GLPP analysis method for fault detection and classification in a dynamic, nonlinear system. In this case study, the normal process dataset is used to determine time lags p and f, set the reduced dimensionality for the GLPP-based CVA model, and fine-tune other parameters. The optimal time lags are identified using the self-correlation function from the summed squares of all measurements. Figure 5.5 shows the variation of the self-correlation coefficient, with confidence bounds of +5% and −5%, and reveals that the optimal time lag is 10.

According to Eq. (5.37), the spectrum radius of these Laplacian matrices is $L_{pglobal} = 33.15$, $L_{plocal} = 21.35$, $L_{fglobal} = 39.29$, $L_{flocal} = 24.14$, respectively. The balance parameter of global and local structure is set as $\eta = 0.6$. The constant parameters in global weight and local weight calculation is set as $\sigma_1 = 48$, $\sigma_2 = 132$ according to experience and trials. The neighborhood number r of data sample is determined by Euclidean norm of sample $\mathbf{y}_i$ and selected as 10. The value of the main parameters in this chapter is listed in Table 5.1.

To evaluate the dynamic performance of the proposed canonical GLPP analysis method, fault detection capabilities are compared with the dynamic GLPP method.

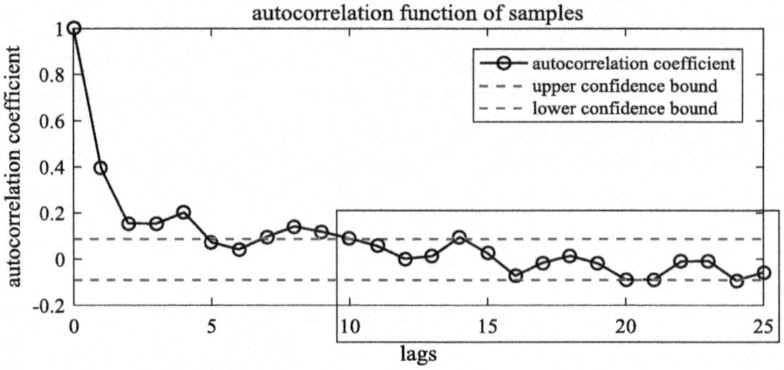

Fig. 5.5 Self-correlation function of samples

Fault detection delay (FDD) and the number of false alarms (NFA) are used as key metrics. FDD is defined as the time interval between the onset of a fault and its detection, measured in samples. NFA refers to the number of samples incorrectly flagged as faults during the first 160 samples of the process operation. Methods with low values for both FDD and NFA are considered to have superior fault detection accuracy and timeliness.

Table 5.2 shows the FDD and NFA of CVA based GLPP method and dynamic GLPP method, number in bold indicates better result. The FDD is reduced a lot through CVA based GLPP apparently, especially with fault 8, fault 18 and fault 20. Both method perform well with low false alarm overall speaking, CVA based GLPP method can detect fault with nearly no false especially, it still make some improvement compared with dynamic GLPP.

Figures 5.6 and 5.7 illustrate the fault detection procedures for the canonical variate analysis-based global-local preserving projection (CVA-GLPP) and the dynamic GLPP methods. In these figures, the red lines represent the control limits calculated using Kernel density estimation, while the blue lines correspond to the T_c^2 and T^2 statistic indices, respectively. The significance level is set to $\alpha = 0.01$. Points marked

Table 5.1 Summary of parameters for the Canonical GLPP analysis algorithm

Parameters	Determined by	Value
The reduced dimension	Akaike Information Criterion	$k = 25$
Lags p and f	Autocorrelation function	$p = f = 10$
Balance parameter of GLPP	Balancing the spectral radius of Laplacian matrices of LPP and GPP model	$\eta = 0.6$
Global weight of GLPP	Experience and trials	$\sigma_1 = 48$
Local weight of GLPP	Experience and trials	$\sigma_2 = 132$
The neighborhood number in GLPP	Euclidean norm	$r = 10$

Table 5.2 Fault detection delay (FDD) and number of false alarm (NFA) comparison

NFA/FDD	Canonical GLPP analysis	Dynamic GLPP
Fault1	**2/0**	7/4
Fault2	17/**0**	**12**/0
Fault5	**0/0**	1/3
Fault6	**0/1**	0/2
Fault7	0/0	0/5
Fault8	**4/0**	23/0
Fault12	**2/11**	3/16
Fault13	**38**/2	38/**0**
Fault17	**20/0**	29/0
Fault18	**57/0**	88/0
Fault20	**78/0**	87/0

with a data cursor indicate the first sample detected as a fault after its occurrence. The ellipses in the figures represent normal samples that were incorrectly classified as faults. For Fault 1, Fault 5, and Fault 7 in Fig. 5.6, the NFA drops to zero with the CVA-GLPP method compared to the dynamic GLPP method. In Fig. 5.7, the FDD for Fault 8 and Fault 17 shows significant reductions, with Fault 8's delay decreasing from 23 samples using the dynamic GLPP method to just 4 samples using CVA-GLPP. These improvements in both FDD and NFA reflect the enhanced fault detection performance of the CVA-based GLPP method.

In summary, the lower FDD and NFA indicate that CVA achieves optimal statistical accuracy for dynamic processes when compared to dynamic methods using extended matrices. This is largely because the past and future observation matrices in the CVA-based GLPP approach encapsulate more dynamic latent information.

Fault identification rate (FIR) is applied to measure the diagnosis accuracy of proposed method, and is defined as the rate of truly classify in the 21 types in this chapter. Since the early 160 samples in test data of 21 fault types are in normal procedure, the remaining 800 samples of all fault types are used to calculate the FIR. this chapter compares canonical GLPP analysis method with CVA-FDA [13], GLPP [14], DGLPP [15] and DPCA [16] method and result shows in Table 5.3, where number in bold indicates the best result.

The proposed method demonstrates the highest identification accuracy across nearly all fault types, with FIR for faults 1, 2, 6, and 7 approaching 100%. Since CVA cannot be directly applied for fault identification, a combined approach using CVA-FDA is introduced for comparison. While CVA-FDA performs reasonably well for several fault types, it struggles significantly with faults 3, 4, 7, 9, 11, 15, 19, and 20. This is largely due to CVA's focus on self-correlation of individual variables, neglecting the cross-correlation among them. In comparisons between DGLPP and GLPP, DGLPP outperforms GLPP, indicating that the dynamic adjustments with the extended matrix enhance fault identification. Conversely, DPCA shows the lowest

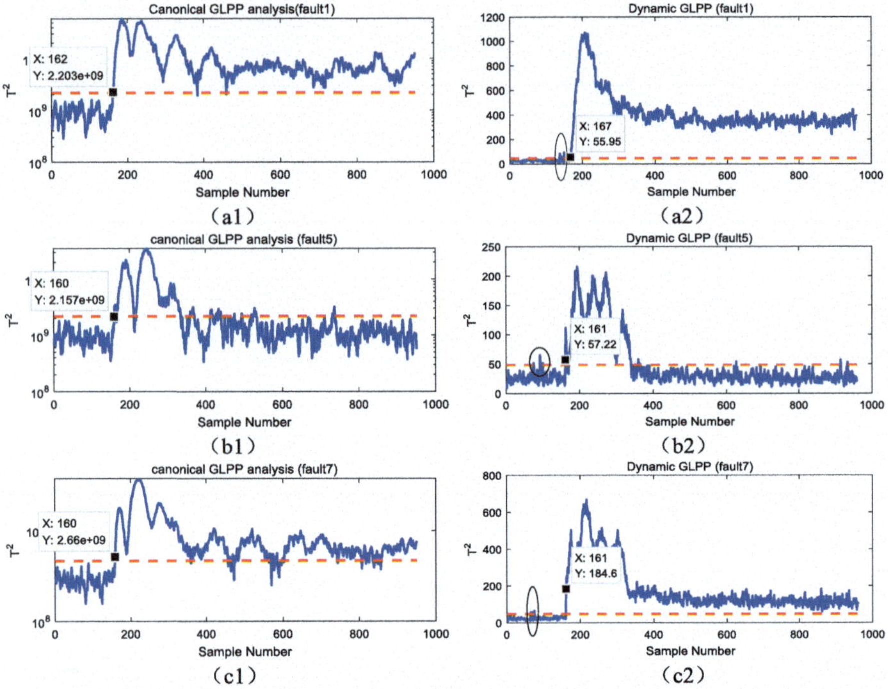

Fig. 5.6 Fault detection of fault 1, fault 5, fault 7

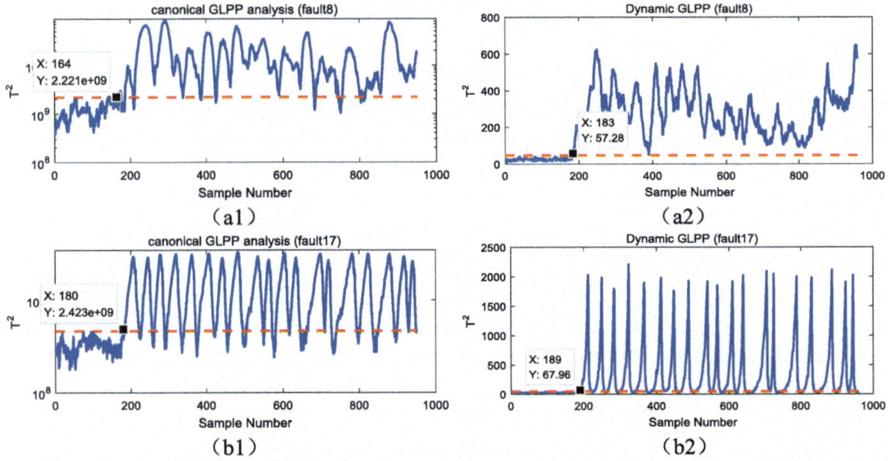

Fig. 5.7 Fault detection of fault 8, fault 17

identification accuracy, as it fails to account for the neighborhood structure, highlighting the importance of both global and local structures in dimensionality reduction for

Table 5.3 Fault identification comparison

Fault type/FIR	Canonical GLPP analysis	CVA-FDA	DGLPP	GLPP	DPCA
Fault#1	**99.75**	94.20	99.25	93.93	95.13
Fault#2	**99.50**	96.70	98.25	95.06	94.25
Fault#3	**20.37**	2.00	11.38	18.45	0
Fault#4	**93.37**	8.22	81.75	92.79	0
Fault#5	79.25	**100**	81.25	43.23	9.25
Fault#6	**100.00**	100	100	85.58	93.50
Fault#7	**100.00**	25.00	81.37	90.39	21.88
Fault#8	88.00	81.00	**96.63**	37.42	43.38
Fault#9	**17.75**	1.20	1.25	17.44	0
Fault#10	27.75	19.80	11.75	**36.66**	1.63
Fault#11	**52.12**	16.40	28.63	30.34	0.50
Fault#12	93.50	**96.70**	93.50	36.53	60.00
Fault#13	62.37	88.70	**92.37**	14.28	74.38
Fault#14	97.75	98.20	86.2	**100**	71.50
Fault#15	**27.87**	11.80	14.00	16.43	0
Fault#16	**39.12**	35.90	6.63	13.52	0
Fault#17	**84.80**	83.40	64.50	84.57	58.13
Fault#18	**89.12**	88.80	88.75	85.20	86.63
Fault#19	**70.50**	6.00	68.25	50.94	0
Fault#20	**62.50**	13.50	29.38	59.03	0.50
Fault#21	**29.80**	27.70	28.00	4.045	0

identification purposes. Overall, the canonical GLPP analysis method exhibits substantial improvements in fault identification due to its consideration of both structural and serial correlations.

5.5 Conclusion

This chapter introduces a novel canonical GLPP analysis model designed for data-driven modeling of dynamic processes. Initially, it constructs an observation matrix with lags based on monitoring samples and derives a Hankel matrix to capture dynamic information. Subsequently, the global and local preserving projection algorithm is applied to the Hankel matrix to create Laplacian matrices for both past and future observations using Cholesky decomposition. The transformation matrix obtained from the objective function of the canonical GLPP algorithm is formulated as a generative singular value decomposition problem. Given that the Hankel matrix encompasses the majority of dynamic variations in the data samples, and the resulting past and future Laplacian matrices are developed from a manifold perspective, this method combines the strengths of CVA in managing high-dimensional data with serial correlation through state space modeling and the benefits of GLPP in pattern recognition. The optimal transformation matrix effectively retains both time series variations and spatial structural information during dimensionality reduction, facilitating timely fault detection and precise identification of dynamically varying data.

Several metrics are introduced based on the proposed model to evaluate its performance. The T_c^2 and Q_c statistics, derived from the canonical GLPP analysis, are designed to monitor the dynamic variations in the process. The parameter selection section provides a detailed explanation of the criteria used to choose the main parameters in this novel model. Additionally, fault detection rate, number of false alarms, and fault identification rate are introduced as key indicators of the monitoring capability.When compared with methods such as CVA-FDA, GLPP, DGLPP, and DPCA, the hybrid CVA-GLPP framework offers a robust and accurate paradigm for dynamic process data monitoring and fault identification. This demonstrates the superiority of the proposed method in terms of both fault detection accuracy and efficiency.

References

1. Jiang B, Braatz RD (2017) Fault detection of process correlation structure using canonical variate analysis-based correlation features. J Process Control 58:131–138
2. Tang Q, Chai Y, Jianfeng Q, Fang X (2019) Industrial process monitoring based on fisher discriminant global-local preserving projection. J Process Control 81:76–86
3. Patricia Odiowei P-E, Cao Y (2009) Nonlinear dynamic process monitoring using canonical variate analysis and kernel density estimations. IEEE Trans Indus Inform 6(1):36–45
4. Larimore WE (1996) Statistical optimality and canonical variate analysis system identification. Signal Process 52(2):131–144
5. Zhang M, Ge Z, Song Z, Ruowei F (2011) Global-local structure analysis model and its application for fault detection and identification. Indus Eng Chem Res 50(11):6837–6848
6. Gonzalez R, Huang B, Lau E (2015) Process monitoring using kernel density estimation and bayesian networking with an industrial case study. ISA Trans 58:330–347
7. Li G, Joe Qin S (2018) Comparative study on monitoring schemes for non-gaussian distributed processes. J Process Control 67:69–82
8. Cao Y, Lei Yu, Deng X, Zhang X (2020) Variable sub-region canonical variate analysis for dynamic process monitoring. IEEE Access 8:37775–37789
9. Ruiz-Cárcel C, Cao Y, Mba D, Lao L, Samuel RT (2015) Statistical process monitoring of a multiphase flow facility. Control Eng Pract 42:74–88
10. Akaike H (1998) Information theory and an extension of the maximum likelihood principle, 199–213
11. Jiang B, Zhu X, Huang D, Paulson JA, Braatz RD (2015) A combined canonical variate analysis and fisher discriminant analysis (cva–fda) approach for fault diagnosis. Comput Chem Eng 77:1–9
12. Larimore WE (1997) Optimal reduced rank modeling, prediction, monitoring and control using canonical variate analysis. IFAC Proc Vol 30(9):61–66
13. Lu Q, Jiang B, Bhushan Gopaluni R, Loewen PD, Braatz RD (2018) Sparse canonical variate analysis approach for process monitoring. J Process Control 71:90–102

14. Bao S, Luo L, Mao J, Tang D (2016) Improved fault detection and diagnosis using sparse global-local preserving projections. J Process Control 47:121–135
15. Tao Y, Shi H, Song B, Tan S (2019) A novel dynamic weight principal component analysis method and hierarchical monitoring strategy for process fault detection and diagnosis. IEEE Trans Indus Electron 67(9):7994–8004
16. Xiao Z, Wang H, Zhou J (2016) Robust dynamic process monitoring based on sparse representation preserving embedding. J Process Control 40:119–133

Chapter 6
Autoencoder Embedded Dictionary Learning for Nonlinear Processes Fault Identification

Industrial processes often exhibit significant nonlinearity due to complex mechanisms, system integrations, and varying operating conditions. While many dictionary learning algorithms have been developed for fault identification, most rely on linear combinations of dictionary atoms, which fail to capture nonlinear relationships, resulting in suboptimal performance. Advances in multilayer neural networks, particularly autoencoders, offer potential solutions but are hindered by the scarcity of fault samples. To address these challenges, this chapter proposes an Autoencoder Embedded Dictionary Learning (AEDL) method for nonlinear industrial fault identification. The approach employs an autoencoder to map linearly inseparable process data into a high-dimensional space, enabling effective dictionary learning. Two supervised graphs leveraging prior process data information are integrated to enhance robustness under limited training samples. The learned dictionary's coding coefficients are then used for fault identification via a simple classifier. Experiments on the Tennessee Eastman process demonstrate that AEDL surpasses existing dictionary learning and nonlinear fault identification methods.

6.1 Motivation

Significant nonlinearity often exists among process data in industrial systems, primarily due to the following factors: (1) Complex physical and chemical reactions create nonlinear relationships between variables. (2) As modern industrial processes become increasingly integrated and complex to meet profit and environmental demands, variable coupling becomes more common. (3) Due to the demands of multiple product grades and varying loads, industrial processes often operate under multiple conditions. This nonlinearity poses challenges for dictionary learning-based fault identification methods, as great deal of these methods model process data using

H. Yin et al., *Data-Driven Fault Diagnosis for Complex Industrial Processes*,
Engineering Applications of Computational Methods 22,
https://doi.org/10.1007/978-981-96-3153-7_6

a linear combination of dictionary atoms. These linear models fall short in capturing the nonlinear relationships among data dimensions, resulting in constrained diagnostic capabilities.

To tackle the issue of nonlinearity, numerous effective techniques have been emerged. A common solution is the use of kernel-based methods, which have been widely employed in recent years [1]. The fundamental concept underlying kernel-based methods entails mapping the original data into a higher-dimensional space using a nonlinear kernel transformation, followed by the application of dictionary learning in this transformed domain [2]. However, selecting an appropriate kernel function is often challenging.

More recently, multilayer neural networks [3–5], known for their robust learning capabilities, have demonstrated superior performance in nonlinear mapping tasks. Among these, the autoencoder is particularly important due to its high performance in data reconstruction [3]. In recent years, dictionary learning has also been integrated into deep learning frameworks. However, these methods often assume the availability of sufficient labeled data for training [6]. In practice, obtaining a large number of labeled data samples for different fault types is challenging. As a result, training on limited faulty data samples can lead to overfitting, which may significantly degrade the performance of these methods [7].

Moreover, industrial process data often contain intrinsic properties such as local structure and label information. These properties represent prior knowledge embedded in the data that can be leveraged to improve diagnostic performance. Extensive experiments suggest that incorporating strong priors can enhance the generalization ability of classification models [8]. From this perspective, considering local and label information is crucial for enhancing fault identification performance, especially when training samples per fault class are limited.

Motivated by the above analysis, this chapter proposes an AEDL approach for nonlinear industrial process fault identification. First, the industrial process data is nonlinearly mapped into a high-dimensional space using the network structure of autoencoders. This nonlinear mapping, driven by a learnable network and loss function, transforms linearly inseparable data into linearly separable data. Subsequently, the dictionary is learned in this high-dimensional space. To address the issue of limited labeled training samples, two supervised graph constraints are introduced into the learning process, which help preserve local structure and enhance discrimination capability. Once the dictionary is learned, new industrial process data can be represented by this dictionary, and the coding coefficients can be used as features for fault identification via a simple classifier.

In summary, the major contribution of this chapter is:

- The proposal of a novel dictionary learning-based fault identification method for nonlinear industrial process systems. This method provides a nonlinear extension of existing dictionary learning techniques, making it more suitable for fault identification in nonlinear industrial processes.
- An autoencoder is employed for nonlinear mapping. Unlike kernel-based approaches, which rely on implicit and predetermined kernel functions, the

autoencoder automatically learns a nonlinear mapping function from the data, effectively capturing the nonlinear structure of industrial process data.
- The learning process incorporates locality structure and label information. By leveraging these priors, the proposed approach enhances the discriminative ability of the features, improving performance even with a limited number of training samples per class.

6.2 Methodology

In this section, the proposed Autoencoder Embedded Dictionary Learning (AEDL) method for industrial process fault identification is described. First, the overall framework of the method is presented. Then, the objective function of the proposed approach is detailed. Furthermore, the specific optimization steps of the method are explained, followed by a discussion of the fault identification model.

6.2.1 *Overall Framework*

The diagram in Fig. 6.1 depicts the comprehensive fault identification framework for industrial processes, which is grounded in dictionary learning. In this framework, an autoencoder is employed to project the original process data onto a high-dimensional feature space. Subsequently, dictionary learning is then applied to uncover the fundamental and intrinsic structures within the process data, leveraging its strengths in data representation. To maximize information extraction, the dictionary and autoencoder are optimized simultaneously during the training stage. Additionally, local geometric structures—which assume that nearby data points should have similar representations—often contain discriminative information about neighboring data pairs. By incorporating locality-constraint priors and label information into the learning process, the proposed approach becomes more robust to variations in training samples. During the testing stage, the learned dictionary and nonlinear mapping function are used to represent the essential features of the process data. The label of new process data can then be predicted using a simple classifier. The framework is divided into three main modules: joint nonlinear mapping and dictionary learning, a supervised graph imposed on nonlinear mapping, and a supervised graph imposed on dictionary learning.

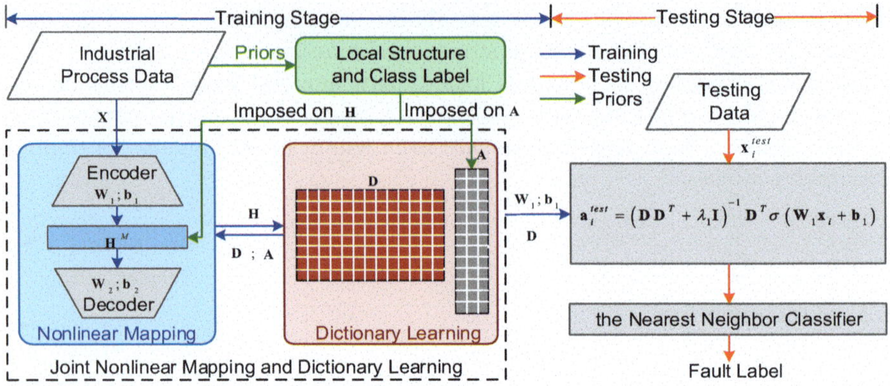

Fig. 6.1 The framework of the proposed AEDL approach

6.2.2 Autoencoder Embedded Dictionary Learning

To address the nonlinear problem, an autoencoder is used to map the process data into a high-dimensional space. Formally, an autoencoder consists of two stages: encoding and decoding. Given the ith original process data $\mathbf{x}_i$, the encoder maps this data into a latent representation $\mathbf{h}_i$ during the encoding stage as follows:

$$\mathbf{h}_i = f\left(\mathbf{W}_1 \cdot \mathbf{x}_i + \mathbf{b}_1\right) \tag{6.1}$$

where $\mathbf{W}_1$ is a $D \times d$ weight matrix, and $\mathbf{b}_1 \in \mathbb{R}^D$ is a bias vector. The latent representation $\mathbf{h}_i$ has d dimensions, with $d < D$. The function $f(\cdot)$ is the sigmoid activation function, defined as $f(x) = \frac{1}{1+\exp(-x)}$. In this way, the autoencoder maps the original process data nonlinearly to a high-dimensional feature space.

In the decoding stage, the decoder reconstructs the process data $\mathbf{x}_i$ from the encoder output $\mathbf{h}_i$ as follows:

$$\hat{\mathbf{x}}_i = f\left(\mathbf{W}_2 \mathbf{h}_i + \mathbf{b}_2\right) \tag{6.2}$$

where $\mathbf{W}_2$ and $\mathbf{b}_2$ denote the weight matrix and bias vector of the decoder, respectively. The autoencoder is optimized by minimizing the difference between the original process data $\mathbf{x}_i$ and the reconstructed data $\hat{x}_i$, calculated as $\left\|\hat{\mathbf{x}}_i - \mathbf{x}_i\right\|_2^2$.

Considering the nonlinear mapping process, the dictionary learning problem can be formulated in the high-dimensional space as follows:

$$\begin{aligned} \min_{\boldsymbol{\Omega},\mathbf{D},\mathbf{A}} J_1 = &\frac{1}{2N}\left(\|\mathbf{H}(\boldsymbol{\Omega}) - \mathbf{DA}\|_F^2 + \left\|\hat{\mathbf{X}}(\boldsymbol{\Omega}) - \mathbf{X}\right\|_F^2\right) \\ &+ \lambda_1\left(\|\mathbf{W}_1\|_F^2 + \|\mathbf{W}_2\|_F^2 + \|\mathbf{A}\|_F^2\right) \end{aligned} \tag{6.3}$$

where $\mathbf{D} \in \mathbb{R}^{D\times K}$ denotes the dictionary learned in the high-dimensional space, and $\mathbf{A} \in \mathbb{R}^{K\times N}$ is the coefficient matrix representing the process data over $\mathbf{D}$. Here, D is the dimension of the high-dimensional space, K is the number of atoms in the learned dictionary, and N is the number of samples.

The parameter set of the autoencoder is denoted as $\mathbf{\Omega} = [\mathbf{W}_1, \mathbf{W}_2, \mathbf{b}_1, \mathbf{b}_2]$. $\mathbf{H}(\mathbf{\Omega}) = [\mathbf{h}_1(\mathbf{\Omega}), \ldots, \mathbf{h}_N(\mathbf{\Omega})]$ represents the hidden representations of the process data, and $\hat{\mathbf{X}}(\mathbf{\Omega}) = \left[\hat{\mathbf{x}}_1(\mathbf{\Omega}), \ldots, \hat{\mathbf{x}}_N(\mathbf{\Omega})\right]$ denotes the reconstructed values of $\mathbf{X}$.

To enhance generalization ability, a weight decay penalty is added to constrain the magnitudes of weights $\mathbf{W}_1$ and $\mathbf{W}_2$. As indicated in Eq. (6.3), the joint nonlinear mapping and dictionary learning module can simultaneously learn a nonlinear mapping function characterized by $\mathbf{\Omega}$ and a dictionary $\mathbf{D}$ within a single optimization framework. This approach efficiently extracts nonlinear features for industrial process fault identification.

According to Eq. (6.3), optimizing the parameters of the autoencoder often requires a large number of training samples. However, in industrial processes, obtaining fault samples can be challenging. Therefore, achieving promising results with a limited number of fault samples is desirable. Fortunately, prior knowledge of the original process data can provide valuable information to improve diagnostic performance. Intuitively, if two points $\mathbf{x}_i$ and $\mathbf{x}_j$ are close in the intrinsic geometry of the data, the corresponding hidden representations $\mathbf{h}_i$ and $\mathbf{h}_j$ in the autoencoder should be similar. For a given sample $\mathbf{x}_i \in \mathbf{X}$, its neighbors are defined as the k nearest data samples, where k is a pre-defined positive constant. To characterize the local structure, the edge weight between two connected data points is determined by the similarity between them as follows:

$$s\left(\mathbf{x}_i, \mathbf{x}_j\right) = \exp\left(-\frac{\left\|\mathbf{x}_i - \mathbf{x}_j\right\|^2}{t}\right) \tag{6.4}$$

where t controls the weight decay speed. The weight matrix $\mathbf{S}_{1ij}$ can then be defined as:

$$\mathbf{S}_{1ij} = \begin{cases} s\left(\mathbf{x}_i, \mathbf{x}_j\right), if\ \mathbf{x}_i \in N\left(\mathbf{x}_j\right) or\ \mathbf{x}_j \in N\left(\mathbf{x}_i\right) \\ 0 \qquad\qquad , otherwise \end{cases} \tag{6.5}$$

where $N(\mathbf{x}_i)$ denotes the k-nearest neighbors of $\mathbf{x}_i$. Given the class labels of the original process data, the weight matrix $\mathbf{S}_{1ij}$ can be further extended to a supervised form as:

$$\mathbf{S}_{1ij} = \begin{cases} s\left(\mathbf{x}_i, \mathbf{x}_j\right) \quad , if\ \mathbf{x}_i \in N\left(\mathbf{x}_j\right) or\ \mathbf{x}_j \in N\left(\mathbf{x}_i\right) \\ \qquad\qquad and\ label\left(\mathbf{x}_i\right) = label\left(\mathbf{x}_j\right) \\ -s\left(\mathbf{x}_i, \mathbf{x}_j\right), if\ \mathbf{x}_i \in N\left(\mathbf{x}_j\right) or\ \mathbf{x}_j \in N\left(\mathbf{x}_i\right) \\ \qquad\qquad and\ label\left(\mathbf{x}_i\right) \neq label\left(\mathbf{x}_j\right) \\ 0 \qquad\qquad , otherwise \end{cases} \tag{6.6}$$

where label$(\mathbf{x}_i)$ denotes the label of $\mathbf{x}_i$. Subsequently, the Laplacian matrices of graph $\mathbf{G}_1$ are calculated as follows:

$$\mathbf{L}_1 = \mathbf{B}_1 - \mathbf{S}_1 \tag{6.7}$$

where $\mathbf{B}_1$ is a diagonal matrix with elements $\mathbf{B}_{1ii} = \sum_{j=1}^{N} \mathbf{S}_{1ij}$. Therefore, the local geometric structure and label information can be incorporated into the discriminative graph as follows:

$$\begin{aligned}\min_{\boldsymbol{\Omega}} J_2 &= \frac{\lambda_2}{2}\left(\sum_{i=1}^{N}\sum_{j=1}^{N}\left\|\mathbf{h}(\boldsymbol{\Omega})_i - \mathbf{h}(\boldsymbol{\Omega})_j\right\|_2^2\right)\mathbf{S}_{1ij} \\ &= \lambda_2 Tr\left[\mathbf{H}(\boldsymbol{\Omega})\mathbf{L}_1\mathbf{H}(\boldsymbol{\Omega})^T\right]\end{aligned} \tag{6.8}$$

where Tr(·) denotes the trace operator, and λ_2 and λ_3 are regularization parameters that penalize the relative contributions. Equation (6.8) transfers both locality relationships and label information of the original process data to the embedded space. This ensures that neighbors with the same label are encouraged to be close in the embedded space, while neighbors from different classes are pushed further apart.

To augment the discriminative power of the acquired dictionary, the coding coefficient matrix **A** is compelled to possess discriminatory characteristics. This is accomplished by introducing a supervised graph during the dictionary learning phase, which minimizes the variance within classes and maximizes the separability between classes of the coding coefficients. Specifically, the supervised graph's weight matrix is formulated as follows:

$$\mathbf{S}_{2ij} = \begin{cases} 1 \ , if\ \mathbf{x}_i \in N(\mathbf{x}_j)\ or\ \mathbf{x}_j \in N(\mathbf{x}_i) \\ \quad and\ label(\mathbf{x}_i) = label(\mathbf{x}_j) \\ -1 \ , if\ \mathbf{x}_i \in N(\mathbf{x}_j)\ or\ \mathbf{x}_j \in N(\mathbf{x}_i) \\ \quad and\ label(\mathbf{x}_i) \neq label(\mathbf{x}_j) \\ 0 \ , otherwise \end{cases} \tag{6.9}$$

Likewise, $N(\mathbf{x}_i)$ denotes the k-nearest neighbors of $\mathbf{x}_i$, and $label(\mathbf{x}_i)$ is the label of $\mathbf{x}_i$. Subsequently, the supervised graph term imposed during the dictionary learning stage is written as:

$$\begin{aligned}\min_{\mathbf{A}} J_3 &= \frac{\lambda_3}{2}\left(\sum_{i=1}^{N}\sum_{j=1}^{N}\left\|\mathbf{A}_i - \mathbf{A}_j\right\|_2^2\right)\mathbf{S}_{2ij} \\ &= \lambda_3 Tr\left(\mathbf{A}\mathbf{L}_2\mathbf{A}^{\mathrm{T}}\right)\end{aligned} \tag{6.10}$$

where $\mathbf{L}_2$ is the Laplacian matrix defined as $\mathbf{L}_2 = \mathbf{B}_2 - \mathbf{S}_2$, and $\mathbf{B}_2$ is a diagonal matrix with elements $\mathbf{B}_{2ii} = \sum_{j=1}^{N} \mathbf{S}_{2ij}$. This supervised graph term in the dictionary learning stage ensures that coding vectors $\mathbf{A}_i$ and $\mathbf{A}_j$ are close to each other if the nearest neighbors $\mathbf{x}_i$ and $\mathbf{x}_j$ belong to the same class, and are maximally separated if $\mathbf{x}_i$ and $\mathbf{x}_j$ are from different classes.

By combining the objective function (6.3) with the supervised graph, the resulting objective function of the proposed approach can be formulated as the following minimization problem:

$$\begin{aligned}\min_{\boldsymbol{\Omega},\mathbf{D},\mathbf{A}} J &= J_1 + J_2 + J_3 \\ &= \frac{1}{2N}\left(\|\mathbf{H}(\boldsymbol{\Omega}) - \mathbf{DA}\|_F^2 + \left\|\hat{X}(\boldsymbol{\Omega}) - \mathbf{X}\right\|_F^2\right) \\ &+ \lambda_1\left(\|\mathbf{W}_1\|_F^2 + \|\mathbf{W}_2\|_F^2 + \|\mathbf{A}\|_F^2\right) \\ &+ \lambda_2 Tr\left[\mathbf{H}(\boldsymbol{\Omega})\mathbf{L}_1\mathbf{H}(\boldsymbol{\Omega})^T\right] \\ &+ \lambda_3 Tr\left(\mathbf{AL}_2\mathbf{A}^T\right)\end{aligned} \tag{6.11}$$

The model consists of four components. The first component ensures that the data **H** in the high-dimensional space can be well represented by the dictionary **D**, while minimizing the reconstruction error between $\hat{\mathbf{X}}(\Omega)$ and **X**. The second component focuses on regularizing the magnitude of the weights and encouraging sparsity during model training. The third and fourth components address the incorporation of locality information with label embedding, applied respectively to the nonlinear mapping and dictionary learning stages. These components are designed to preserve local structure and enhance class discrimination. The parameters λ_1, λ_2, and λ_3 are positive regularization constants. The objective function demonstrates that the proposed approach can simultaneously learn a nonlinear mapping function $f(\cdot)$ characterized by $\boldsymbol{\Omega}$ and a dictionary **D**. This integrated optimization method effectively extracts nonlinear and discriminative features for industrial process fault identification tasks.

6.2.3 *Iterative Optimization*

In this section, we detail the steps to learn the parameter set $\boldsymbol{\Omega}$, the dictionary **D**, and the coding coefficient matrix **A**. The objective function (6.11) is divided into three sub-problems, which are optimized iteratively by updating one variable while keeping the others fixed. The optimization procedures are described in the following subsections.

Learn $\boldsymbol{\Omega}$ with fixed D andA

When **D** and **A** are fixed, (6.11) can be rewritten as

$$\begin{aligned}\min_{\boldsymbol{\Omega}} J = &\frac{1}{2N}\|\mathbf{H}(\boldsymbol{\Omega}) - \mathbf{DA}\|_F^2 \\ &+ \frac{1}{2N}\left\|\hat{X}(\boldsymbol{\Omega}) - \mathbf{X}\right\|_F^2 \\ &+ \lambda_1\left(\|\mathbf{W}_1\|_F^2 + \|\mathbf{W}_2\|_F^2\right) \\ &+ \lambda_2 Tr\left[\mathbf{H}(\boldsymbol{\Omega})\mathbf{L}_1\mathbf{H}(\boldsymbol{\Omega})^T\right]\end{aligned} \tag{6.12}$$

The back-propagation algorithm and the Limited Memory Broyden–Fletcher–Goldfarb–Shanno (L-BFGS) algorithm [9] are used to update $\mathbf{\Omega}$. The key step in this process involves computing the partial derivatives $\partial J/\partial \mathbf{W}_1$, $\partial J/\partial \mathbf{b}_1$, $\partial J/\partial \mathbf{W}_2$, and $\partial J/\partial \mathbf{b}_2$. The details are given as:

$$\begin{aligned}\frac{\partial J}{\partial \mathbf{W}_1} &= \frac{1}{N}\left[(\mathbf{H}-\mathbf{DA}) \odot \mathbf{H} \odot (\mathbf{1}-\mathbf{H})\,\mathbf{X}^T + \mathbf{\Delta}_1 \mathbf{X}^T\right] \\ &\quad + 2\lambda_1 \mathbf{W}_1 + 2\lambda_2 \left[\mathbf{H}\mathbf{L}_1 \odot \mathbf{H} \odot (\mathbf{1}-\mathbf{H})\right]\mathbf{X}^T\end{aligned} \tag{6.13}$$

$$\begin{aligned}\frac{\partial J}{\partial \mathbf{b}_1} &= \frac{1}{N}\left[(\mathbf{H}-\mathbf{DA}) \odot \mathbf{H} \odot (\mathbf{1}-\mathbf{H})\,\mathbf{1} + \mathbf{\Delta}_1 \mathbf{1}\right] \\ &\quad + 2\lambda_2 \left[\mathbf{H}\mathbf{L}_1 \odot \mathbf{H} \odot (\mathbf{1}-\mathbf{H})\right]\mathbf{1}\end{aligned} \tag{6.14}$$

$$\frac{\partial J}{\partial \mathbf{W}_2} = \frac{1}{N}\mathbf{\Delta}_2 \mathbf{H}^T + 2\lambda_1 \mathbf{W}_2 \tag{6.15}$$

$$\frac{\partial J}{\partial \mathbf{b}_2} = \frac{1}{N}\mathbf{\Delta}_2 \mathbf{1} \tag{6.16}$$

where $\odot$ denotes the element-wise product operator. $\mathbf{\Delta}_1$ are $\mathbf{\Delta}_2$ computed by

$$\mathbf{\Delta}_1 = \left(\mathbf{W}_2^T \mathbf{\Delta}_2\right) \odot \mathbf{H} \odot (\mathbf{1}-\mathbf{H}) \tag{6.17}$$

$$\mathbf{\Delta}_2 = \left(\hat{\mathbf{X}} - \mathbf{X}\right) \odot \hat{\mathbf{X}} \odot \left(\mathbf{1} - \hat{\mathbf{X}}\right) \tag{6.18}$$

To simplify notation, $\mathbf{H}(\mathbf{\Omega})$ and $\hat{\mathbf{X}}(\mathbf{\Omega})$ are simplified to $\mathbf{H}$ and $\hat{\mathbf{X}}$, respectively. Then $\mathbf{\Omega}$ can be updated as follows until convergence:

$$\begin{cases}\mathbf{W}_i = \mathbf{W}_i - \eta\,\partial J/\partial \mathbf{W}_i \\ \mathbf{b}_i = \mathbf{b}_i - \eta\,\partial J/\partial \mathbf{b}_i\end{cases} \tag{6.19}$$

where η is the learning rate controlling the convergence speed, $i = 1, 2$.

Learn A with fixed $\mathbf{\Omega}$ and D

When $\mathbf{\Omega}$ and $\mathbf{D}$ are fixed, (6.11) can be rewritten as

$$\min_{\mathbf{A}} J = \|\mathbf{H}(\mathbf{\Omega}) - \mathbf{DA}\|_F^2 + \lambda_1 \|\mathbf{A}\|_F^2 + \lambda_3 Tr\left(\mathbf{A}\mathbf{L}_2\mathbf{A}^T\right) \tag{6.20}$$

The coefficient matrix $\mathbf{A}$ is optimized column-wise. Specifically, each coefficient vector $\mathbf{a}_i$ is updated while keeping the other coefficients $\mathbf{a}_j$ (where $j \neq i$) fixed. The optimization of each $\mathbf{a}_i$ can be simplified as follows:

$$\min_{\mathbf{a}_i} J = \|\mathbf{h}_i - \mathbf{D}\mathbf{a}_i\|_2^2 + \lambda_1 \|\mathbf{a}_i\|_2^2 + \lambda_3 f(\mathbf{a}_i) \tag{6.21}$$

where $f(\mathbf{a}_i) = 2\left[\mathbf{a}_i^T(\mathbf{A}\mathbf{L}_{2i}) - \mathbf{a}_i^T\mathbf{a}_i\mathbf{L}_{2ii}\right]$, $\mathbf{L}_{2i}$ denotes the ith column of matrix $\mathbf{L}_2$ and $\mathbf{L}_{2ii}$ denotes the (i, i)-element of matrix $\mathbf{L}_2$. The analytic solution can be obtained by setting the first order derivative of (6.21) to zero as follows:

$$\begin{aligned}\mathbf{a}_i^* &= \left[\mathbf{D}^T\mathbf{D} + (\lambda_1 + \lambda_3\mathbf{L}_{ii})\mathbf{I}\right]^{-1} \\ &\times \left(\mathbf{D}^T\mathbf{h}_i - \lambda_3 \sum_{m\neq i} \mathbf{a}_m\mathbf{L}_{2mi}\right)\end{aligned} \tag{6.22}$$

Learn D with fixed Ω and A

When $\boldsymbol{\Omega}$ and $\mathbf{A}$ are fixed, (6.22) can be rewritten as

$$\min_{\mathbf{D}} \|\mathbf{H}(\boldsymbol{\Omega}) - \mathbf{D}\mathbf{A}\|_F^2 \quad s.t. \quad \|\mathbf{D}_i\|_2^2 = 1 \tag{6.23}$$

where $\|\mathbf{D}_i\|_2^2 = 1$ ensures that each $\mathbf{D}_i$ is a unit vector, enhancing the stability of the learned dictionary, with $i = 1, \ldots, K$. This formulation results in a quadratic programming problem, which can be effectively solved using the Lagrange multiplier method described in [10]. The dictionary $\mathbf{D}$ can then be obtained as follows:

$$\mathbf{D} = \mathbf{H}(\boldsymbol{\Omega})\mathbf{A}^T\left(\mathbf{A}\mathbf{A}^T + \boldsymbol{\Gamma}\right)^{-1} \tag{6.24}$$

where $\boldsymbol{\Gamma} \in \mathbb{R}^{K\times K}$ is a diagonal matrix with elements defined as $\boldsymbol{\Gamma}_{ii} = \gamma_i$ for $i = 1, \ldots, K$, where γ_i represents the Lagrange multiplier for the i-th equality constraint. These steps are iteratively repeated until convergence is achieved or the maximum number of iterations is reached.

6.2.4 Fault Identification

Once the nonlinear mapping function $f(\cdot)$ characterized by $\boldsymbol{\Omega}$ and the dictionary $\mathbf{D}$ are learned, the fault identification task can be performed as follows. For a test sample $\mathbf{x}_i^{test}$, it can be encoded as $\mathbf{a}_i^{test}$ using Eq. (6.25).

$$\mathbf{a}_i^{test} = \left(\mathbf{D}\mathbf{D}^T + \lambda_1\mathbf{I}\right)^{-1}\mathbf{D}^T f(\mathbf{W}_1\mathbf{h}_i + \mathbf{b}_1) \tag{6.25}$$

Then, the label of $\mathbf{a}_i^{test}$ can be predicted using a simple classifier. In this chapter, the Nearest Neighbor Classifier is employed for its simplicity and efficiency. Specifically, the nearest neighbor rule assigns $\mathbf{a}_i^{test}$ to the class j ($j \in \{1, 2, \ldots, C\}$) based on the following rule:

If $i_0 = \arg\min_i \left\|\mathbf{a}_i^{test} - \mathbf{a}_i\right\|$ and $label\,(\mathbf{a}_{i0}) = j$, then $label\left(\mathbf{a}_i^{test}\right) = j$

6.2.5 Theoretical Analysis

The computational cost of the proposed approach arises from several parts. First, the time complexity for calculating the similarity values for a single sample $\mathbf{x}_i$ in the dataset $\mathbf{X} \in \mathbb{R}^{d \times N}$ is $O(dN)$, where N is the number of samples and d is the data dimensionality. Consequently, the computational complexity for constructing the k-nearest neighbor (kNN) graph for the entire training dataset is $O(dN^2)$. When the dictionary $\mathbf{D}$ and the coefficient matrix $\mathbf{A}$ are fixed, the time complexity of using the L-BFGS algorithm to optimize Eq. (6.11) is $O(tdDN^2)$, where t is the number of iterations of the L-BFGS algorithm and D is the hidden layer size of the autoencoder. When $\boldsymbol{\Omega}$ and $\mathbf{A}$ are fixed, the time complexity of updating the dictionary $\mathbf{D}$ is $O(K^3DN)$, where K is the number of dictionary atoms. Similarly, when $\boldsymbol{\Omega}$ and $\mathbf{D}$ are fixed, the time complexity of updating the coefficient matrix $\mathbf{A}$ is also $O(K^3DN)$. Therefore, the overall time complexity of the proposed approach is $O\left(T\left(tdDN^2 + K^3DN\right)\right)$, where T is the maximum number of iterations of the proposed approach.

The optimization procedure of the proposed approach involves three subproblems formulated in Eqs. (6.12), (6.20), and (6.23). In the first subproblem, the L-BFGS algorithm is employed to optimize Eq. (6.12) with respect to $\boldsymbol{\Omega}$. Given that the L-BFGS algorithm is convergent [9], the optimization process for $\boldsymbol{\Omega}$ will also converge. In the second subproblem, Eq. (6.20) is convex and differentiable with respect to each column of $\mathbf{a_i}$, ensuring that the objective function value decreases after solving for $\mathbf{A}$ in each iteration. For the third subproblem, Eq. (6.23) is convex and differentiable with respect to $\mathbf{D}$. The optimal dictionary can be obtained using the closed-form solution in Eq. (6.24), ensuring that the objective function value decreases after solving for $\mathbf{D}$ in each iteration. Thus, the proposed approach is iteratively descent, meaning that each optimization step progressively reduces the objective function value.

The convergence characteristics of the proposed approach are demonstrated through experimentation, with the convergence curve shown in Fig. 6.2. In this figure, the x-axis represents the iteration number, while the y-axis shows the value of the objective function. The observed trend indicates that the objective function values decrease and stabilize at a fixed value after approximately 15 iterations. This rapid convergence underscores the efficiency of the proposed approach for fault identification tasks, making it highly suitable for practical applications. Therefore, a maximum of 15 iterations is set for all experiments to ensure both effectiveness and efficiency.

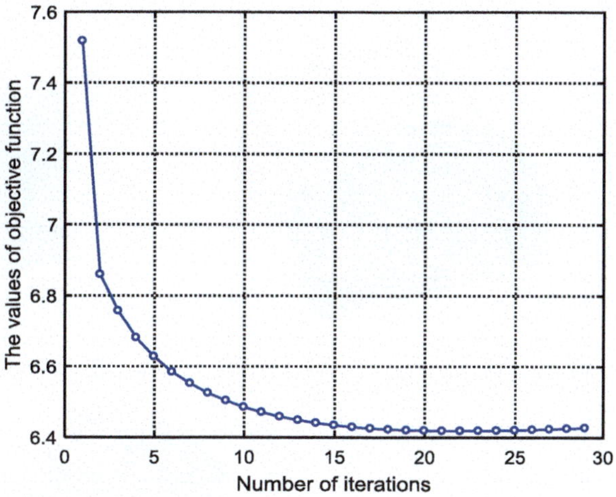

Fig. 6.2 Convergence behavior of the proposed method on the TEP dataset

6.3 Case Study

6.3.1 Tennessee Eastman Process

In this section, the proposed approach is evaluated using the TEP dataset, a widely recognized simulation program known for its nonlinear relationships between variables. As noted by Zhao et al. [7], faults 9, 10, 11, 14, 16, and 21 are particularly challenging to diagnose and are thus excluded from the experiments. For the remaining 15 faults, each fault is represented by 480 randomly selected training samples and 800 test samples.

The initial comparison of the proposed method involves five dictionary learning-based techniques: K-Singular Value Decomposition (K-SVD) [11], Fisher Discrimination Dictionary Learning (FDDL) [12], Support Vector Guided Dictionary Learning (SVGDL) [13], Kernel K-SVD (KKSVD) [2], and Linearized Kernel Dictionary Learning (LKDL) [14]. In addition, the proposed approach is evaluated against several nonlinear methodologies, namely Exponential Discriminant Analysis (EDA) [15], Sparse Exponential Discriminant Analysis (SEDA) [7], a one-hidden-layer autoencoder (Autoencoder-I), and a stacked autoencoder featuring two hidden layers (Autoencoder-II).

The fault identification performance is evaluated using classification accuracy as the index. Higher accuracy values indicate better performance of the fault identification method. The diagnostic accuracy index is defined as follows:

$$Acc(i) = 100 \times {N_i}^T/N_i;\ Ave = \sum_{i=1}^{C} N_i \times Acc(i)/N \tag{6.26}$$

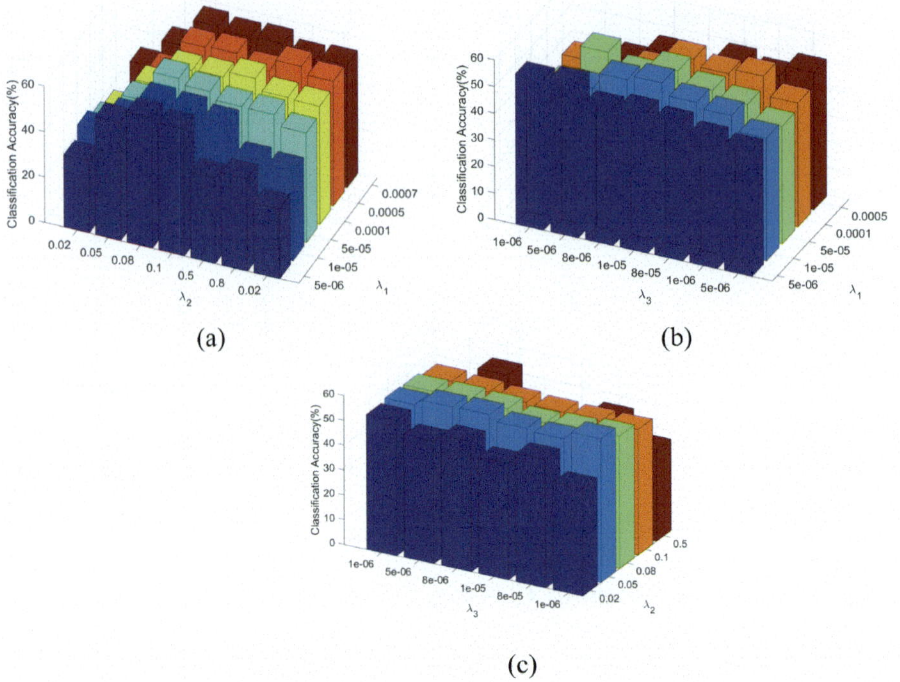

Fig. 6.3 Parameters selection under different parameter configurations on the TEP dataset. **a** effects of tuning λ_1 and λ_2 on the performance over fixing λ_3, **b** effects of tuning λ_1 and λ_3 on the performance over fixing λ_2, and **c** effects of tuning λ_2 and λ_3 on the performance over fixing λ_1

where $Acc(i)$ denotes the diagnostic accuracy of Class i, and Ave represents the average diagnostic accuracy. N_i^T and N_i denote the number of correctly classified samples and the total number of samples in Class i, respectively. N represents the total number of samples in the dataset.

Parameter Settings

Note that the three key parameters, λ_1, λ_2, and λ_3, significantly impact the performance of the proposed approach. A grid search method was employed to determine the values of these parameters by fixing some and tuning others. Specifically, λ_1 was tuned from the set $[5 \times 10^{-3}, 1 \times 10^{-5}, 5 \times 10^{-5}, 1 \times 10^{-4}, 5 \times 10^{-4}, 7 \times 10^{-4}, 1 \times 10^{-3}]$, while λ_2 and λ_3 were tuned from the sets $[0.02, 0.05, 0.1, 0.5, 0.8]$ and $[1 \times 10^{-6}, 5 \times 10^{-6}, 8 \times 10^{-6}, 1 \times 10^{-5}, 8 \times 10^{-5}]$, respectively.

The effects of the parameters λ_1 and λ_2 are first studied with λ_3 fixed at 5×10^{-6}. The fault identification results with respect to λ_1 and λ_2 are reported in Fig. 6.3a. The optimal result is achieved when $\lambda_1 = 5 \times 10^{-5}$ and $\lambda_2 = 8 \times 10^{-2}$, indicating that setting λ_1 and λ_2 to these values is empirically reasonable. Similarly, when λ_2 is set to 8×10^{-2}, the optimal result is attained with $\lambda_1 = 5 \times 10^{-5}$ and $\lambda_3 = 5 \times 10^{-6}$.

It is observed that the proposed method performs well across a wide range of λ_2 and λ_3 when λ_1 is fixed. Therefore, λ_1, λ_2, and λ_3 are set to 5×10^{-5}, 8×10^{-2}, and 5×10^{-6}, respectively, in the experiment.

In the proposed method, the number of nearest neighbors k, the dimensionality of the autoencoder's hidden layer D, and the count of dictionary atoms K are also crucial parameters in the proposed method. These parameters are determined using a rigorous five-fold cross-validation strategy. Specifically, the training samples are randomly partitioned into five subsets, with four subsets serving as training and the remaining subset used for validation. This procedure is iterated five times, with each subset used once as validation data. The final results from the five folds are averaged to produce a single result, and the parameters yielding the best fault identification performance are selected for the testing sets. Given the multiple parameters to be tuned, they can be optimized iteratively by updating one parameter while keeping others fixed.

The number of nearest neighbors k is varied from 3 to 8, with the experimental results shown in Fig. 6.4. The maximum fault identification accuracy is achieved when $k = 6$; accuracy declines as k increases beyond this point. This observation is intuitive, as a higher number of nearest neighbors captures more local information, but too many neighbors can lead to overlap between samples from different categories. This suggests that an optimal k value is around 6. Therefore, the number of nearest neighbors k is set to 6 in the experiment.

The effect of different hidden layer sizes on fault identification performance was evaluated for sizes ranging from 60 to 140. As shown in Fig. 6.5, the fault identification accuracy improves progressively with an increase in the hidden layer size, achieving the best performance at approximately 110. This improvement is due to the increased capacity of network parameters with larger hidden layers. However, if the hidden layer size becomes too large, the proposed approach may suffer from

Fig. 6.4 Fault identification results on different datasets with varying k

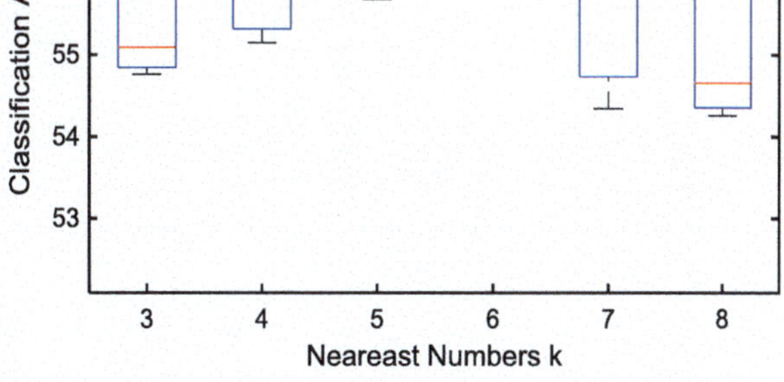

overfitting, which can degrade classification performance when training samples are limited. Therefore, the hidden layer size of the autoencoder is empirically set to 110 for the experiment.

The number of dictionary atoms was varied from 105 to 195. The fault identification results are shown in Fig. 6.6, indicating that increasing the number of dictionary atoms generally enhances the diagnostic performance of the proposed approach. The performance stabilizes when the number of dictionary atoms reaches approximately 165. Beyond this point, there is no significant improvement or decline in performance. Therefore, the number of dictionary atoms K is set to 165 for the experiment. Additionally, the proposed approach was compared with other methods across different numbers of dictionary atoms. The results suggest that the proposed approach consistently achieves the best performance compared to the competing methods at each dictionary size and maintains relatively strong performance even at lower dictionary sizes. These observations confirm that accounting for the nonlinearity of industrial process data while preserving local and discriminative structures benefits the final fault identification performance.

Comparisons with Dictionary Learning Approaches

Initially, the proposed approach is benchmarked against recent methodologies in discriminative dictionary learning and kernel-based dictionary learning, utilizing the KSVD method as the foundational baseline. Discriminative dictionary learning methods, which incorporate discriminant power in the dictionary learning process, include FDDL [12] and SVGDL [13]. However, these methods are based on linear models. Kernel-based dictionary learning methods include KKSVD [2] and LKDL [14].

Both kernel-based methods rely on predefined kernel functions. To ensure a fair and consistent comparison, the same Gaussian kernel is used in the experiment. The diagnostic results on the testing data are summarized in Table 6.1, with the best result for each fault highlighted in bold.

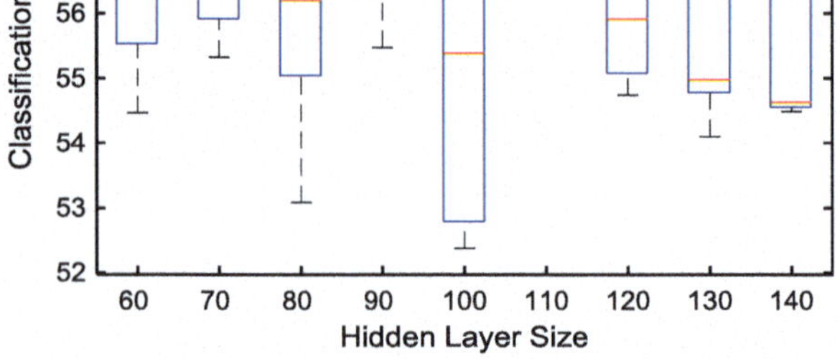

Fig. 6.5 Fault identification results with different hidden layers

Fig. 6.6 Fault identification results with different dictionary sizes

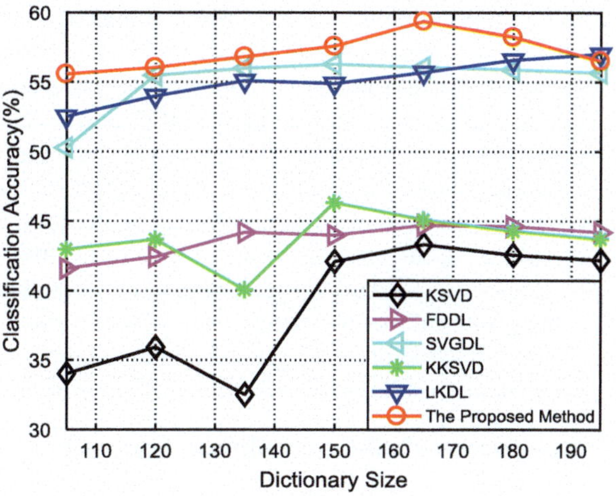

Table 6.1 shows the following observations:

(1) The proposed approach significantly outperforms the KSVD method, showing an average improvement of 19.28%. This enhancement is primarily due to the KSVD method's lack of consideration for nonlinear and discriminative structures within the TEP data. The proposed approach leverages nonlinearity and discrimination to achieve superior performance.

Table 6.1 Fault identification accuracy (%) on the testing data with different methods for selected faults (All but Fault 9, 10, 11 , 14, 16 and 21)

Faults	Dictionary learning approaches					Other nonlinear approaches				IDL	AEDL
	KSVD	FDDL	SVGDL	KKSVD	LKDL	EDA	SEDA	AE-I	AE-II		
Fault 1	92.12	92.38	97.75	95.88	97.38	80.94	81.46	97.13	96.00	98.13	**98.63**
Fault 2	95.25	93.50	96.25	96.88	96.13	77.92	80.42	96.13	94.88	96.00	**97.13**
Fault 3	**32.50**	22.38	25.25	29.88	29.25	15.31	11.46	23.25	20.13	23.50	29.38
Fault 4	45.25	28.63	43.75	56.88	39.00	13.54	20.21	30.88	26.63	42.63	**79.00**
Fault 5	12.38	**82.13**	31.25	4.38	17.63	29.38	35.63	22.88	20.63	35.25	36.75
Fault 6	90.63	49.63	98.63	88.75	97.38	76.25	77.19	98.38	98.38	**98.75**	98.63
Fault 7	44.13	96.50	98.25	83.13	87.63	68.65	76.98	82.38	70.63	98.00	**99.50**
Fault 8	9.50	39.88	28.00	6.75	33.75	18.44	29.27	26.25	22.88	29.63	**41.50**
Fault 12	5.38	31.00	36.88	28.63	45.63	20.52	20.21	33.13	31.13	38.13	**35.50**
Fault 13	**48.88**	28.63	32.00	1.63	30.25	13.33	25.62	25.63	24.63	34.38	31.88
Fault 15	25.38	2.50	26.75	2.63	13.25	21.56	13.44	23.38	20.13	25.88	**30.63**
Fault 17	46.75	64.75	42.00	**72.88**	45.88	43.75	46.98	30.63	25.63	44.13	46.00
Fault 18	2.88	45.25	81.25	72.63	79.38	63.12	62.40	80.38	79.50	81.38	**84.25**
Fault 19	11.00	11.63	52.75	6.13	36.50	20.83	21.67	49.88	48.75	48.50	**53.13**
Fault 20	**69.38**	54.13	53.5	47.00	58.38	29.69	34.79	46.50	45.25	49.88	58.63
Ave	42.09	49.53	56.28	46.27	53.83	39.55	42.51	52.12	48.35	56.28	**61.37**

(2) The proposed approach achieves the highest average fault identification result (61.37%) compared to FDDL (49.53%) and SVGDL (56.28%). These findings suggest that the proposed approach is more effective in diagnosing faults in industrial processes. Although FDDL and SVGDL can extract discriminative information from process data, they remain linear techniques and may overlook nonlinear structures, leading to limited performance.
(3) When compared with kernel-based methods, the proposed approach demonstrates a significant improvement in fault identification performance. This highlights that a standard kernel function does not always guarantee optimal results. The proposed method, which employs an autoencoder for automatic nonlinear mapping while preserving local and discriminative structures, provides the best diagnostic results across the fifteen faults considered.

Comparisons with Other Nonlinear Approaches

In previous sections, the proposed approach was compared with three types of dictionary learning methods. However, there are other fault identification techniques that effectively capture the nonlinear structures in industrial processes. For instance, Exponential Discriminant Analysis (EDA) [15] transforms scatter matrices into a new space rather than mapping the original data. Additionally, Zhao et al. [7] developed Sparse Exponential Discriminant Analysis (SEDA), which transforms the original data into a new space using distance diffusion mapping and automatically selects key variables responsible for faults.

Given that the proposed method uses an autoencoder for nonlinear mapping, we also compared it with autoencoder-based methods. Specifically, we considered a single-layer autoencoder (Autoencoder-I) and a stacked autoencoder with two hidden layers (Autoencoder-II). The hidden layer sizes for Autoencoder-I and Autoencoder-II are 165 and 105 for the first layer and 165 for the second layer, respectively. The final identification results on the testing data are shown in Table 6.1. For context, the results for EDA and SEDA are taken from [7].

As shown in Table 6.1, the proposed approach demonstrates significant improvement compared to EDA and SEDA, indicating its effectiveness in addressing the nonlinearity problem in industrial process fault identification. The proposed method also outperforms both Autoencoder-I and Autoencoder-II. This is primarily because autoencoder-based methods do not incorporate discriminative information, limiting the effectiveness of deep learning-based methods on such datasets.

Lastly, we assess the significance of combining nonlinear mapping and dictionary learning. The proposed method is compared with an Individual Nonlinear Mapping and Dictionary Learning (IDL) method, where nonlinear mapping and dictionary learning are performed separately. Table 6.1 presents the fault identification results, showing that the proposed approach significantly outperforms the IDL method. This improvement is attributed to the fact that separating nonlinear mapping and dictionary learning may result in the loss of useful information. The joint learning framework of the proposed method enhances its performance.

Regularization Terms

The contributions of different regularization terms to the proposed approach are investigated by evaluating classification performance with and without the J_2 and J_3 terms on selected faults. Specifically, setting $\lambda_2 = 0$ removes the J_2 term from model (6.11) to assess its impact during the nonlinear mapping stage. Similarly, setting $\lambda_3 = 0$ removes the J_3 term to evaluate its significance during the dictionary learning stage. When both λ_2 and λ_3 are zero, neither local geometrical structure nor discriminative information is preserved.

Additionally, the fault identification results of the proposed method are examined against varying numbers of training samples, ranging from 100 to 450 samples per fault.

Figure 6.7 presents the average fault identification results with different regularization terms and numbers of training samples. The results show a significant drop in performance when J_2 and J_3 are removed from the model, regardless of the number of training samples. Furthermore, J_2 contributes more to the final identification performance than J_3 for each number of training samples. This is because the dictionary is learned directly on data after nonlinear mapping, and preserving locality and label information during the nonlinear mapping stage benefits the subsequent dictionary learning stage. Clearly, the best fault identification results are achieved when both J_2 and J_3 terms are included in the model. This indicates that both terms are crucial for optimal fault identification performance.

Regarding the average fault identification accuracy with different numbers of training samples, Fig. 6.7 shows that while identification performance declines with fewer training samples, the proposed approach with both J_2 and J_3 terms maintains satisfactory performance. Notably, with only 100 training samples per class, the proposed approach achieves an accuracy of 47.10%, whereas the best result from other methods is only 34.00%. This demonstrates that the proposed approach offers relatively stable performance even with limited training samples, thanks to the incorporation of prior knowledge within the process data.

6.4 Conclusion

This chapter presents an autoencoder-embedded dictionary learning approach designed to address the nonlinear characteristics of industrial process data for fault identification. The proposed method automatically learns a mapping function that projects industrial process data into a high-dimensional space, effectively exploiting the nonlinear structure to enhance diagnostic performance. By simultaneously optimizing nonlinear mapping and dictionary learning within a joint framework, the approach preserves more information, further improving diagnostic accuracy. Additionally, incorporating locality and category label information allows the method to enhance fault identification performance even with a limited number of training samples per class.

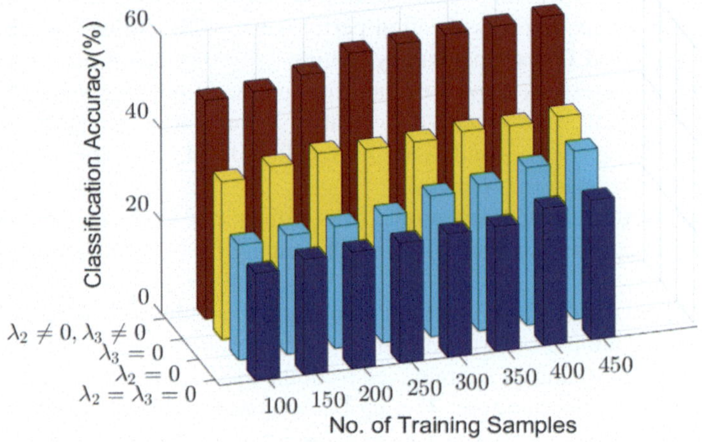

Fig. 6.7 Fault identification results with different regularization terms are evaluated using varying numbers of training samples per class

References

1. Zhang L, Lin J, Karim R (2018) Adaptive kernel density-based anomaly detection for nonlinear systems. Knowl Based Syst 139:50–63
2. Van Nguyen H, Patel VM, Nasrabadi NM, Chellappa R (2013) Design of non-linear kernel dictionaries for object recognition. IEEE Trans Image Process 22(12):5123–5135
3. Ma S, Chen M, Jianwen W, Wang Y, Jia B, Jiang Y (2019) High-voltage circuit breaker fault diagnosis using a hybrid feature transformation approach based on random forest and stacked autoencoder. IEEE Trans Indus Electron 66(12):9777–9788
4. Yang D, Pang Y, Zhou B, Li K (2019) Fault diagnosis for energy internet using correlation processing-based convolutional neural networks. IEEE Trans Syst Man Cybern Syst 49(8):1739–1748
5. Wanke Yu, Zhao C (2020) Broad convolutional neural network based industrial process fault diagnosis with incremental learning capability. IEEE Trans Indus Electron 67(6):5081–5091
6. Lei Y, Yang B, Jiang X, Jia F, Li N, Nandi AK (2020) Applications of machine learning to machine fault diagnosis: a review and roadmap. Mech Syst Signal Process 138:106587
7. Wanke Yu, Zhao C (2018) Sparse exponential discriminant analysis and its application to fault diagnosis. IEEE Trans Indus Electron 65(7):5931–5940
8. Foroughi H, Ray N, Zhang H (2017) Object classification with joint projection and low-rank dictionary learning. IEEE Trans Image Process 27(2):806–821
9. Liu DC, Nocedal J (1989) On the limited memory bfgs method for large scale optimization. Mathem Programm 45(1):503–528
10. Lee H, Battle A, Raina R, Ng AY (2007) Efficient sparse coding algorithms. In: Advances in neural information processing systems, pp 801–808
11. Aharon M, Elad M, Bruckstein A (2006) K-svd: an algorithm for designing overcomplete dictionaries for sparse representation. IEEE Trans Signal Process 54(11):4311–4322
12. Yang M, Zhang L, Feng X, Zhang D (2014) Sparse representation based fisher discrimination dictionary learning for image classification. Int J Comput Vis 109(3):209–232
13. Cai S, Zuo W, Zhang L, Feng X, Wang P (2014) Support vector guided dictionary learning. In: European conference on computer vision, pp 624–639. Springer, Cham

14. Golts A, Elad M (2016) Linearized kernel dictionary learning. IEEE J Sel Topics Signal Process 10(4):726–739
15. Zhang T, Fang B, Tang YY, Shang Z, Xu B (2010) Generalized discriminant analysis: a matrix exponential approach. IEEE Trans Syst Man Cybern Part B (Cybernetics) 40(1):186–197

Chapter 7
Multi-grained Matrix Factorization for Processes Fault Identification Under Multi-mode Working Conditions

Industrial processes often operate under diverse conditions to fulfill manufacturing strategies. A common approach involves segmenting data based on operational modes and developing tailored learning models for each. However, data from the same mode can still exhibit complex patterns due to non-Gaussian behaviors and fault occurrences. Merely analyzing coarse-grained correlations between modes may insufficiently partition modes. Addressing this, this chapter introduces a novel fault identification method for multi-mode industrial processes. Initially, a hierarchical clustering strategy captures the multi-grained information within process data, modeling correlations both across modes (different operation conditions) and within modes (patterns in each mode). Subsequently, a feature learning algorithm leveraging non-negative matrix factorization (NMF) is introduced to extract data features, enabling samples to be represented by the identified multi-grained structural information. A weighted metric is also designed to accurately measure the feature similarities obtained through the NMF. Notably, our framework employs an ℓ_p-norm ($0 < p \leq 1$) minimization strategy to deal with noise and outliers in practical processes. The efficacy of the proposed framework is demonstrated through a numerical example and the multi-phase flow process.

7.1 Motivation

Industrial processes frequently operate under varying conditions due to environmental changes, material variations, equipment degradation, and economic factors. The multimodal and nonlinear nature of these processes poses significant challenges to achieving high-accuracy fault identification. A prevalent approach for modeling multimode industrial processes involves mode identification [1]. Initially, this method segregates multimode process data into distinct sub-groups, (usually according to the mode correlation), and subsequently constructs a unique model tailored to each

H. Yin et al., *Data-Driven Fault Diagnosis for Complex Industrial Processes*,
Engineering Applications of Computational Methods 22,
https://doi.org/10.1007/978-981-96-3153-7_7

mode or several localized models that fit the respective operational regions. A common strategy for model partition is unsupervised grouping via clustering algorithms. Numerous studies, for instance, use Gaussian Mixture Model (GMM) [2] based clustering for mode identification. However, Gaussian distribution assumption in GMM limits its efficacy in addressing non-Gaussian characteristics prevalent in practical industrial processes. Additionally, its performance is highly susceptible to noise and outliers [3], a limitation shared by K-Means (KM) and Fuzzy C-Means (FCM) based partition methods [4] can uncover non-hyperspherical data structures. Notable examples include the Fast Search and Find of Density Peaks algorithm (FSFDP) [5] and neighborhood-information-based local density [6] estimation. Some works [7] also employ spectral clustering for mode identification, but its computational demands become prohibitive for extremely large datasets, thereby restricting its widespread application in practical cases.

Achieving successful multi-mode process modeling necessitates the precise identification of data structures. Yet, accurately partition modes remains difficult due to the disruptions caused by faults in real-world multi-mode industrial processes, as illustrated in Fig. 7.1. Samples belonging to the same mode (indicated by the colorful circles) exhibit similar underlying behaviors, whereas samples from distinct modes display differing properties, forming groups with varied means and covariances. In the presence of faults (represented by gray crosses), group structures are prone to exhibiting hyper-elliptical and non-convex patterns, contributing to the multi-grained nature of process data. As previously mentioned, traditional partition methods simply divide data based on the number of modes, enabling them to uncover only the structures between modes, i.e., coarse-granulation, while failing to explore the patterns

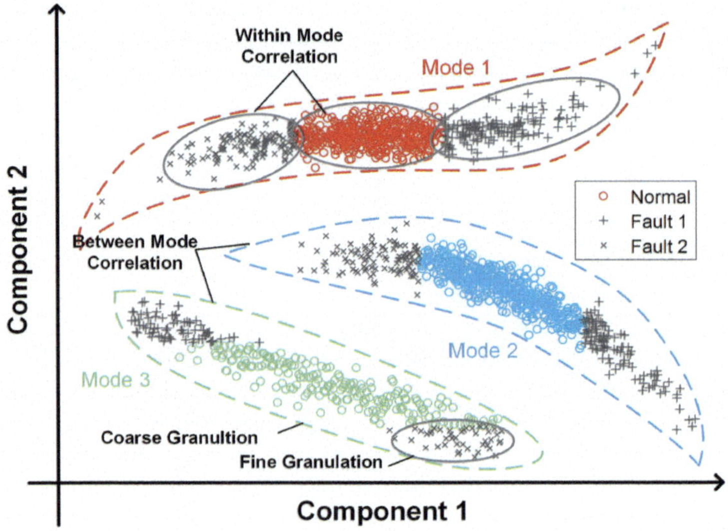

Fig. 7.1 Example visualization of multimode industrial process data with hyper-ellipse structures

within modes, i.e., fine-granulation. Although some approaches [8, 9] can achieve fine-grained partition by increasing the number of Gaussian components in the GMM, they solely focus on within-mode correlations and neglect the higher-level descriptions between modes. Hence, it is imperative to develop effective partition strategies that can simultaneously detect both coarse-grained and fine-grained structures within multi-mode process data.

Drawing from the aforementioned observations, this chapter introduces a multi-grained clustering algorithm termed the Robust K-Multiple-Means algorithm (RKMM), designed to segment multi-mode process data. By leveraging a hierarchical clustering approach, RKMM facilitates the discovery of both coarse-grained (modes) and fine-grained (patterns) structures, enabling precise mode modeling. Additionally, a Non-negative Matrix Factorization (NMF) based method is proposed to extract the inherent data features associated with the identified granulation correlations. Specifically, each sample can be characterized by its membership in both coarse-grained modes and fine-grained patterns, demonstrating robust expressive power in capturing fault characteristics. To enhance the suitability of the similarity metric for fault identification, a weighted metric is designed to reasonably quantify membership similarity. As a result, incoming samples can be accurately classified based on their implicit relationships with historical data. Furthermore, the overall framework adopts a ℓ_p-norm for objective function reconstruction, which is capable of accommodating substantial reconstruction errors, thereby ensuring robustness against noise and outliers. The contributions of this chapter can be summarized as follows:

- This chapter presents a multi-grained clustering technique referred to as RKMM. This method's hierarchical structure facilitates the simultaneous identification of both coarse-grained and fine-grained characteristics within collected industrial data, accurately modeling the correlations both between and within modes of multi-mode processes.
- This chapter presents a fault feature learning algorithm based on RNMF. The algorithm facilitates the representation of each sample through a nonnegative combination of identified fine-grained and coarse-grained information, thereby enabling the learned features to provide a clear interpretation of both operating conditions and fault categories.
- This chapter outlines a minimization strategy based on the ℓ_p-norm for reconstructing the objective function. This strategy enhances the model's robustness, enabling it to tolerate noise and outliers in practical process data.

7.2 Methodology

7.2.1 Robust K-Multiple-Mean Based Mode Partition

Mode partition is usually viewed as a clustering problem. It divides industrial process data into several sub-groups in an unsupervised manner and each group is regarded as an operation mode. K-Means is intensively adopted tools because of its simpleness and efficiency.

Consider a set of data collected from an industrial process, denoted as $\mathbf{X} = [\mathbf{x}_1, ..., \mathbf{x}_N] \in \Re^{d \times N}$. The partition goal of K-Means is to divide data $\mathbf{X}$ into K modes. Thus, the objective function can be formulated as follow:

$$\min_{\mathbf{S},\mathbf{P}} \sum_{i}^{N} \sum_{j}^{K} s_{ij} \left\| \mathbf{x}_i - \mathbf{p}_j \right\|_2 \quad s.t. \ \mathbf{S} \geq 0 \tag{7.1}$$

where p_j represents the center of the j-th cluster. s_{ij} signifies the membership degree of the i-th data within j-th mode. The squared error criterion employed by K-Means clustering is effective when dealing with hyper-spherical data structures. Unfortunately, practical datasets often exhibit hyper-elliptical shapes or even non-convex patterns (non-Gaussian), as shown in Fig. 7.1. One option could be increase the number of clusters K in K-Means, but this approach neglects the between-mode information entirely. Consequently, this framework adopts a multi-grained clustering approach for mode partition.

As shown in Fig. 7.2, our objective in partition is to categorize N data points into K coarse-grained modes, each containing M fine-grained patterns, where $K < M < N$. Initially, we focus solely on clustering the data into M fine-grained patterns (from the bottom layer to the middle layer in Fig. 7.2. Consequently, the objective function can be expressed as follows:

$$\min_{\mathbf{S},\mathbf{P}} \sum_{i}^{N} \sum_{j}^{M} s_{ij} \left\| \mathbf{x}_i - \mathbf{p}_j \right\|_2 + \gamma \left\| \mathbf{S} \right\|_F^2 \quad s.t. \ \mathbf{S} \geq 0 \tag{7.2}$$

This bears a close resemblance to the K-means algorithm presented in Eq. (7.1). Here, p_j represents the center of the j-th pattern. s_{ij} indicates the membership degree of the i-th data point in the j-th pattern. γ is a positive parameter that serves to regulate both the smoothness and sparsity of the matrix $\mathbf{S}$.

Furthermore, our method necessitates the aggregation of M patterns into K modes to learn the mode-level information (from the middle layer to the top layer in Fig. 7.2. To address this challenge, we introduce a constraint. Chung [10] observes that when $rank(\tilde{L}_S) = N + M - K$, the bipartite graph corresponding to the nonnegative matrix $\mathbf{S}$ comprises K connected components (namely, K clusters). In this constraint, $\tilde{L}_S$ is the normalized Laplacian matrix associated with $\mathbf{S}$, $\tilde{L}_S = \mathbf{I} - \mathbf{D}^{1/2} \begin{bmatrix} & \mathbf{S} \\ \mathbf{S}^T & \end{bmatrix} \mathbf{D}^{1/2}$

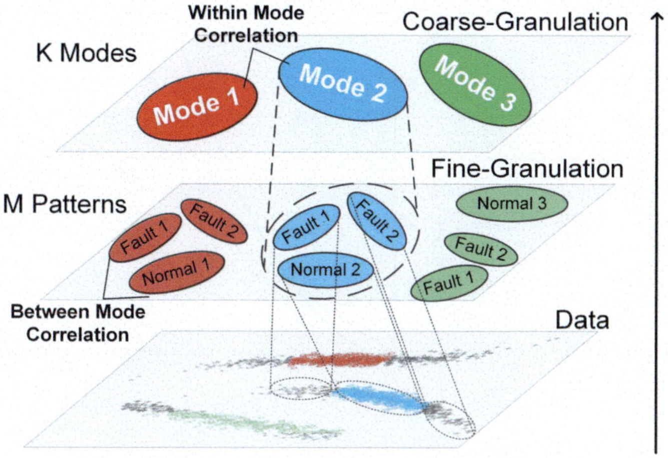

Fig. 7.2 An illustration of K-Multiple-Mean for coarse-granulation and fine-granulation discovering

where **I** is the identity matrix and **D** is a diagonal matrix whose $d_{ii} = \sum_j \begin{bmatrix} & \mathbf{S} \\ \mathbf{S}^T & \end{bmatrix}_{ij}$. Leveraging this theorem, Eq. (7.2) can be utilized to group M patterns into K modes. Consequently, the K-Multiple-Mean based partition algorithm is formulated as outlined below:

$$\min_{\mathbf{S},\mathbf{P}} \sum_{i}^{N} \sum_{j}^{M} s_{ij} \left\| \mathbf{x}_i - \mathbf{p}_j \right\|_2 + \gamma \left\| \mathbf{S} \right\|_F^2 \\ s.t.\ \mathbf{S} \geq 0,\ rank(\tilde{L}_S) = N + M - K \tag{7.3}$$

Note that, the reconstruction error in the first term of Eq. (7.3) is measured using the ℓ_2-norm. Figure 7.3 illustrates the variation in loss as the parameter p changes. Specifically, when $p = 2$ in the ℓ_p-norm, a significant reconstruction error leads to a substantial increase in the objective function's loss. In other words, the squared distance makes outliers in practical industrial processes easily dominate the optimization process. Consequently, models based on the ℓ_2-norm have limited tolerance for contaminated samples in practical processes. Conversely, because $p \in (0, 1]$ can mitigate the impact of large reconstruction errors, this chapter employs a more robust ℓ_p-norm ($p \in (0, 1]$) for the RKMM. By modifying Eq. (7.3), a robust partition version can be formulated as follows:

$$\min_{\mathbf{S},\mathbf{P}} \sum_{i}^{N} \sum_{j}^{M} s_{ij} \left\| \mathbf{x}_i - \mathbf{p}_j \right\|_p + \gamma \left\| \mathbf{S} \right\|_F^2 \\ s.t.\ \mathbf{S} \geq 0,\ rank(\tilde{L}_S) = N + M - K \tag{7.4}$$

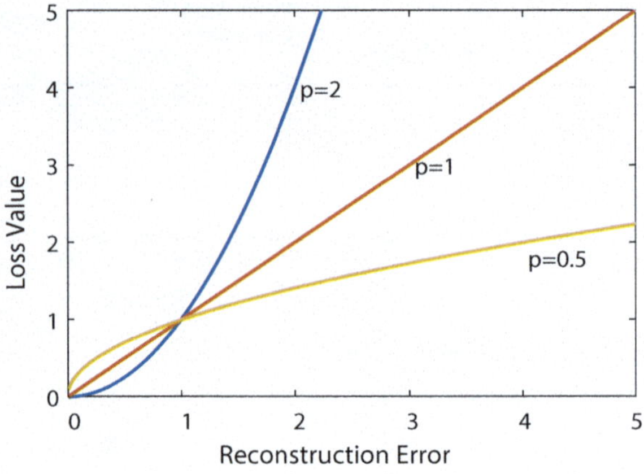

Fig. 7.3 Reconstruction error versus loss value with difference ℓ_p-norm

Here, $\| \|_p$ is the ℓ_p-norm for a vector to measure the reconstruction error:

$$\|\mathbf{a}\|_p = (\sum_i |a_i|^p)^{1/p} \tag{7.5}$$

.

The equation labeled as Eq. (7.4) can be addressed through an alternating optimization approach, i.e., updating one factor while keeping others fixed [11]. In comparison to previous partition methods, our framework's robust K-Multiple-Mean offers several advantages: (1) it is capable of uncovering multi-grained data structures, thereby facilitating accurate mode partitioning; (2) it exhibits enhanced robustness owing to the incorporation of the ℓ_p-norm; (3) it is able to detect non-Gaussian, and even non-convex, distributions within multi-mode process data.

7.2.2 *Robust Non-negative Matrix Factorization Based Fault Feature Learning*

After partition, each sample is anticipated to be characterized by the membership within the identified modes and patterns. As illustrated in Fig. 7.4, the basis matrix $\mathbf{W} \in \Re^{d\times M}$ in Eq. (7.4) is initialized using the extracted pattern centers $[p_1, ..., p_M] \in \Re^{d\times M}$. Due to the parts-based representation capability of Non-Negative Matrix Factorization, samples can be represented as a linear combination of patterns through the derived feature matrix $\mathbf{H} \in \Re^{M\times N}$.

In the conventional Non-Negative Matrix Factorization framework, the Frobenius-norm, because of its reliance on squared distances, tends to be susceptible to noise and outliers. Consequently, following the approach outlined in the preceding section, we employ a $\ell_{2,p}$-norm ($p \in (0, 1]$) minimization strategy to enhance robustness. Thus,

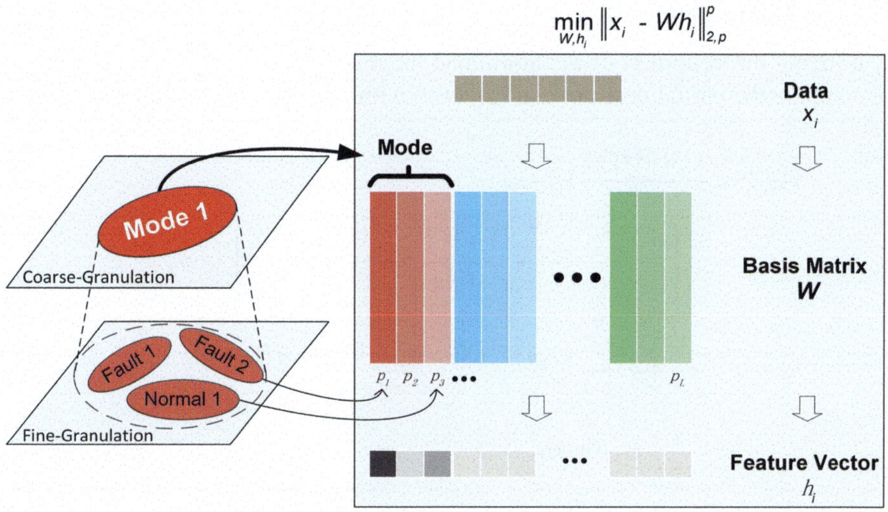

Fig. 7.4 An illustration of Non-Negative Matrix Factorization based fault feature learning

the formulation of our proposed Robust Non-Negative Matrix Factorization (RNMF) model is written as follows:

$$\min_{\mathbf{W},\mathbf{H}} \quad \frac{1}{2}\|\mathbf{X}-\mathbf{W}\mathbf{H}\|_{2,p}^{p} \quad s.t. \mathbf{W}, \mathbf{H} \geq 0 \tag{7.6}$$

Each row $\mathbf{h}_i$ of $\mathbf{H}$ corresponds to the feature of the ith sample. Here, $\|\,\|_{2,p}$ denotes the $\ell_{2,p}$-norm for a matrix:

$$\|\mathbf{A}\|_{2,p} = (\sum_{i} \|\mathbf{a}_{i\cdot}\|_2^p)^{1/p}. \tag{7.7}$$

To solve this non-convex problem, we develop an iterative multiplicative algorithm. Letting $\mathbf{\Theta} = \mathbf{X} - \mathbf{WH}$, the objective function in Eq. (7.6) can rewritten as follows.

$$\mathcal{J} = \frac{1}{2}\|\mathbf{\Theta}\|_{2,p}^{p} \tag{7.8}$$

Given the non-negativity constraints on $\mathbf{W}$ and $\mathbf{H}$, Lagrangian multipliers Φ, Ψ are introduced to enforce the conditions $\mathbf{W} \geq 0$ and $\mathbf{H} \geq 0$, respectively. Then, the Lagrangian function corresponding to Eq. (7.8) is constructed as follows:

$$\mathcal{L}(\mathbf{W}, \mathbf{H}, \mathbf{\Phi}, \mathbf{\Psi}) = \frac{1}{2}\|\mathbf{\Theta}\|_{2,p}^{p} + Tr(\mathbf{\Phi}\mathbf{W}^T) + Tr(\mathbf{\Psi}\mathbf{H}^T) \tag{7.9}$$

Fixing H and updating W

By utilizing the definition of $\ell_{2,p}$-norm and the matrix property $\|\mathbf{A}\|_F = Tr(\mathbf{A}^T\mathbf{A})$, we compute the partial derivative of Eq. (7.9) with respect to $\mathbf{W}$ as outlined below:

$$
\begin{aligned}
\nabla_{\mathbf{W}}\mathcal{L} &= \frac{1}{2}\left(\frac{\partial\,\|\boldsymbol{\Theta}\|_{2,p}^{p}}{\partial W_{ih}}\right) + \phi_{ih} \\
&= \frac{1}{2}\left\{\frac{\partial \sum_{n=1}^{N}(\sum_{j=1}^{M}(X_{jn} - \sum_{s=1}^{M} W_{js}H_{sn})^2)^{\frac{p}{2}}}{\partial W_{ih}}\right\} + \phi_{ih} \\
&= -\frac{p}{2}\sum_{n=1}^{N}(\sum_{j=1}^{M}\Theta_{jn}^2)^{\frac{p}{2}-1}(X_{in} - \sum_{s=1}^{M} W_{is}H_{sn})H_{hn} + \phi_{ih} \\
&= -\sum_{n=1}^{N} D_{nn}\Theta_{in}H_{hn} + \phi_{ih}
\end{aligned}
\tag{7.10}
$$

where $D_{nn} = \frac{p}{2\|\boldsymbol{\Theta}_n\|_2^{2-p}}$.

Setting diagonal matrix $\mathbf{D} = [D_{nn}]$, the Eq. (7.10) can be rewritten as:

$$
\begin{aligned}
\nabla_{\mathbf{W}}\mathcal{L} &= -\Theta\mathbf{D}\mathbf{H}^T + \boldsymbol{\Phi} \\
&= \mathbf{WHDH}^T - \mathbf{XDH}^T + \boldsymbol{\Phi}
\end{aligned}
\tag{7.11}
$$

Setting Eq. (7.11) be 0 and use the Kuhn-Tucker condition $\phi_{ih}W_{ih} = 0$, we can obtain the following equations:

$$
[\mathbf{WHDH}^T - \mathbf{XDH}^T]_{ih}W_{ih} = 0 \tag{7.12}
$$

Then, the multiplicative updating rule for the basis matrix $\mathbf{W}$ is written as follow:

$$
W_{ih} \leftarrow W_{ih}\frac{[\mathbf{XDH}^T]_{ih}}{[\mathbf{WHDH}^T]_{ih}} \tag{7.13}
$$

Fixing W and updating H

Similarly, the partial derivative of Eq. (7.9) with respect to $\mathbf{H}$ is calculated as follow:

$$
\begin{aligned}
\nabla_{\mathbf{H}}\mathcal{L} &= -\mathbf{W}^T\Theta\mathbf{D} + \Psi \\
&= \mathbf{W}^T(\mathbf{WH} - \mathbf{X})\mathbf{D} + \Psi
\end{aligned}
\tag{7.14}
$$

Setting Eq. (7.14) be 0 and using the Kuhn-Tucker condition $\psi_{hj}H_{hj} = 0$, we can obtain the following equation:

$$
[\mathbf{W}^T\mathbf{WHD} - \mathbf{W}^T\mathbf{XD}]_{hj}H_{hj} = 0 \tag{7.15}
$$

Algorithm 1 Updating procedure of the proposed $\ell_{2,p}$ Non-Negative Matrix Factorization.

1: Initialize $\mathbf{H}^0 \in \Re_+^{M\times N}$ by the cluster indicator obtained by the RKMM based algorithm;
2: Initialize $\mathbf{W}^0 \in \Re_+^{d\times M}$ by the **P** obtained by the RKMM based algorithm;
3: Set $\mathbf{D}^0 \in \Re_+^{N\times N}$ as identity matrix;
4: **repeat**
5: Update $\mathbf{H}^{t+1}$ by Eq. (7.16);
6: Update $\mathbf{W}^{t+1}$ by Eq. (7.13);
7: $\mathbf{\Theta}^{t+1} = \mathbf{X} - \mathbf{W}^{t+1}\mathbf{H}^{t+1}$;
8: $\mathbf{D}^{t+1} = \begin{pmatrix} \frac{p}{2\left\|\mathbf{\Theta}_1^{t+1}\right\|_2^{2-p}} & & \\ & \ddots & \\ & & \frac{p}{2\left\|\mathbf{\Theta}_N^{t+1}\right\|_2^{2-p}} \end{pmatrix}$
9: $t = t + 1$
10: **until** the stop criterion is reached
11: Obtain the final **H**;

Subsequently, the multiplicative updating rule for the feature matrix **H** is written as follow:

$$H_{hj} \leftarrow H_{hj} \frac{[\mathbf{W}^T\mathbf{X}\mathbf{D}]_{hj}}{[\mathbf{W}^T\mathbf{W}\mathbf{H}\mathbf{D}]_{hj}} \tag{7.16}$$

The procedure of the RNMF based feature learning algorithm is outlined in Algorithm 1. The iteration stop in step 10 criterion could be either achieving the predefined maximum number of iterations (e.g. $MaxIter = 100$) or reaching a sufficiently small loss variation ($|\mathcal{O}_{t-1} - \mathcal{O}_t|/\mathcal{O}_{t-1} < \epsilon$, where $\mathcal{O}_t$ is the loss function value in the t-th iteration).

7.2.3 *Fault Identification*

We utilize the matrix **H** for fault identification purposes. Specifically, each row of **H** represents the i-th sample, denoted as $\mathbf{h}_i$.

The task of fault identification can be framed as a classification problem, where each sample feature $\mathbf{h}_i$ is assigned a fault class label y_i. To accomplish this, various classifiers can be considered. In this chapter, we adopt the K-Nearest Neighbor (KNN) classifier due to its simplicity and effectiveness. In practice, for a given test sample $\mathbf{h}^*$, KNN identifies its k nearest neighbors, denoted as $\mathbf{N}_k(\mathbf{h}^*)$ from the training set and assigns labels based on these neighbors:

$$y^* = arg \max_{y^*} \sum_{\mathbf{h}^* \in \mathbf{N}_k(\mathbf{h}^*)} \mathbb{I}(y^* = y_i) \tag{7.17}$$

where $\mathbb{I}$ is an indicator function. If $y^* = y_i$, then $\mathbb{I} = 1$.

However, the use of the Euclidean distance as the metric for determining the top k nearest samples in set $\mathbf{N}_k$ is inappropriate for our framework. To illustrate this point more effectively, an example is provided in Fig. 7.5. After modeling with the RNMF, each sample can be characterized by a feature based on its membership in various modes and patterns. Consider four samples belonging to patterns A, B, C and D, with features $[1, 0, 0, 0]$, $[0, 1, 0, 0]$, $[0, 0, 1, 0]$ and $[0, 0, 0, 1]$ respectively. When their similarities are assessed using the standard Euclidean distance, all pairwise distances are found to be $\sqrt{2}$. This outcome is clearly unintuitive, as the samples belong to distinct classes.

7.2.4 The Overall Framework

The workflow of our proposed fault identification model is delineated in Fig. 7.6. Notably, both the training and testing datasets are processed concurrently within the mode partitioning module and feature extraction module. Then, these datasets are split for training and evaluating the weighted KNN classifier. Once the model has been thoroughly trained, for any newly acquired samples, we employ the matrix reconstruction formula presented in Eq. (7.18) to obtain their features and feed them to the KNN classifier for fault identification.

Remarks: New Samples

When a novel sample $\mathbf{x}^n$ arrives, we can leverage the already obtained robust NMF model to extract its feature representation $\mathbf{h}^n$. This can be achieved via matrix reconstruction:

$$\mathbf{h}^n \approx \mathbf{W}^{\dagger}\mathbf{x}^n \tag{7.18}$$

where $\dagger$ denotes the Pseudo-inverse.

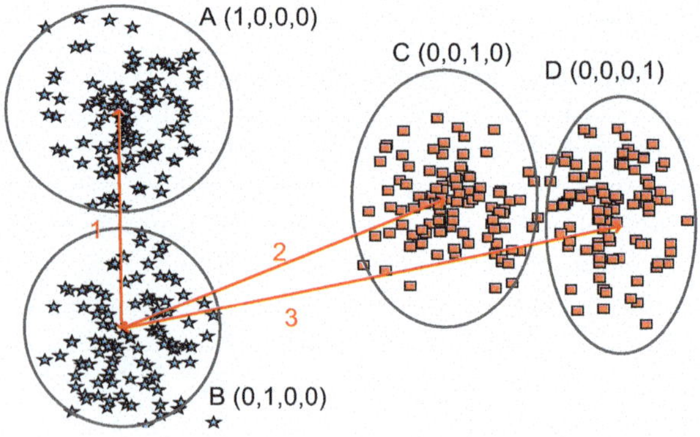

Fig. 7.5 Distance measurements among different patterns

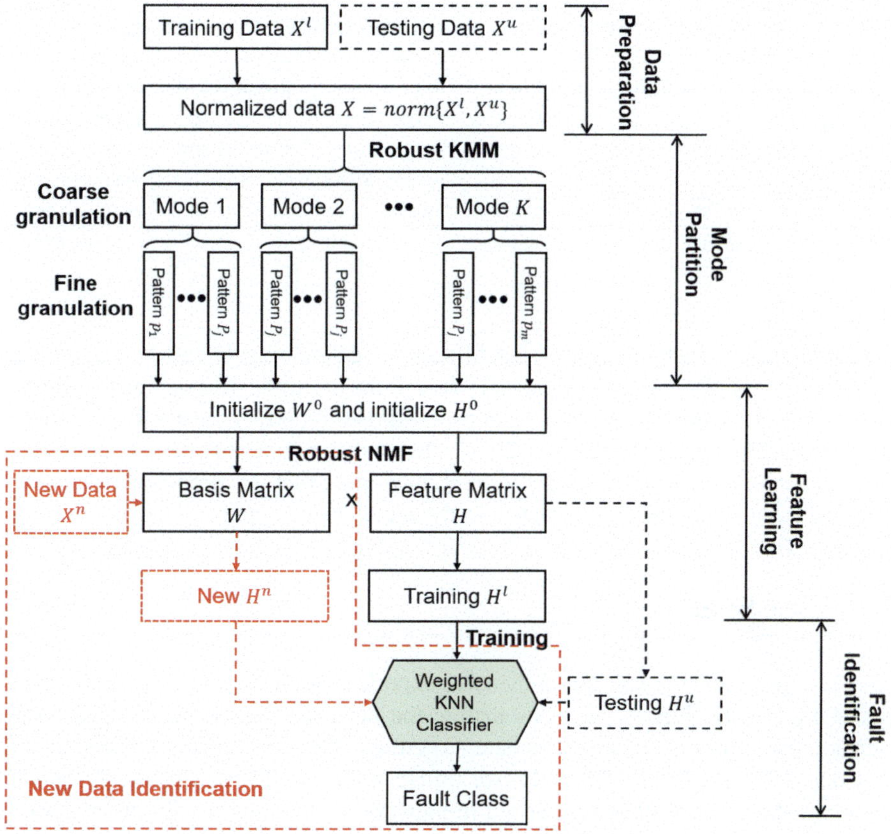

Fig. 7.6 The workflow of the proposed framework for multimode process fault identification

Remarks: Parameter Selection

In the RKMM, two essential parameters must be predefined: the number of modes K and the number of patterns M. The choice of these parameters is crucial, yet currently, there is no definitive mathematical criterion to guide their selection. In this regard, insights into the physical behavior of industrial systems can offer valuable guidance. Precisely, the parameter K can be determined based on the number of operational conditions, whereas the parameter M corresponds to the aggregate of class counts across each operational condition.

Remarks: Computational Complexity

We further analyze the computational complexity associated with our proposed model. Initially, in RKMM, computing $\mathbf{S}$ as per Eq. (7.3) necessitates $\mathcal{O}(N(dM + KM + M^2))$ operations and involves t_1 sub-iterations. Updating $\mathbf{P}$ in Eq. (7.3) requires $\mathcal{O}(dMNt_2)$ operations. Therefore, the total computational cost of RKMM amounts to $\mathcal{O}(N((dM + KM + M^2)t_1 + dM)t_2)$. For the robust NMF, updating $\mathbf{W}$

Table 7.1 The clustering results of different partition algorithms in terms of accuracy

Datasets	KM	FCM	GMM	FSFDP	RKMM
Two Moons	85.42%±0.08%	84.82%±0.40%	91.01%±0.00%	**100.00%±0.00%**	**97.97%±1.31%**
Two Rings	50.86%±0.42%	50.69%±0.44%	75.00%±0.00%	71.30%±0.00%	**100.00%±0.00%**
Smile	40.06%±2.15%	38.76%±2.59%	93.30%±0.00%	51.90%±0.00%	**97.59%±9.06%**
Iris	–	–	**96.67%±0.00%**	64.67%±0.00%	**96.88%±1.05%**
Ecoli	54.75%±3.98%	52.32%±4.00%	**78.87%±0.00%**	46.13%±0.00%	**78.18%±2.29%**
USPS	60.65%±6.06%	27.35%±3.96%	16.10%±0.00%	–	**69.29%±4.87%**

and **H** entails $\mathcal{O}(dMN)$ operations. Additionally, calculating $\boldsymbol{\Theta}$ and **D** necessitates $\mathcal{O}(dMN)$ and $\mathcal{O}(dN)$ operations, respectively. If the multiplicative updates terminate after t_3 iterations, the overall cost of robust NMF is $\mathcal{O}(t_3 dMN)$. Lastly, the computational cost of the weighted KNN is $\mathcal{O}(kN)$. It is worth noting that the computational cost of each module scales linearly with N, indicating that our proposed framework exhibits linear scalability with respect to the number of samples.

7.3 Case Study

In this section, we provide comprehensive experimental results to validate our proposed framework. Firstly, we demonstrate the performance of the RKMM based partition algorithm in discovering data structures. Then, we apply our framework to a numerical illustration. Finally, we utilize a real-world multimode process to showcase the efficacy of our proposed framework, contrasting it with other leading methodologies in the state-of-the-art.

Firstly, 6 public datasets[1] are utilized to validate the rationality of the RKMM via clustering tasks.

The benchmark methods include several prevalent partition methods: K-Means (KM) [4, 12], Fuzzy C-Means (FCM) [13], Gaussian Mixture Model (GMM) [2, 14] and Fast Search and Find of Density Peaks algorithm (FSFDP) [5]. In this section, we set parameters K and C of KM and FCM to match the actual number of clusters. Similarly, the number of Gaussian components in GMM is set to the true cluster count. The setting of FSFDP follows the suggestion in its original work. For our RKMM, we utilize the values of K and M as suggested in the previous section. The clustering performance is assessed using accuracy. To avoid randomness, we conduct each experiment 100 times and compute the mean and standard deviation of the results.

The clustering outcomes are presented in Table 7.1, where the optimal results are highlighted in bold, and unacceptable results are denoted by a dash '-'. As can be seen from this table, the proposed RKMM consistently outperforms other algorithms in terms of accuracy.

[1] http://archive.ics.uci.edu/ml/index.php.

To gain insights into how the RKMM clusters data at both fine and coarse granularities, we visualize the clustering results on the Two Moons dataset. As illustrated in Fig. 7.7, the yellow stars in each figure represent the clusters/patterns centers. When K and C are set to the true number of clusters (2 in this case), both KM and FCM attempt to discover the coarse-grained relationship and fail to identify the two-moon structure, as shown in Fig. 7.7a, b. If K increased (20 in this case), as shown in Fig. 7.7c, while KM is able to detect the fine-grained clusters (encircled by the black boundaries), it fails to capture the overall structure. In contrast, Fig. 7.7d demonstrates that the proposed RKMM partition algorithm accurately groups the data by leveraging both fine and coarse granularities. These observations indicate that the RKMM-based partition algorithm can identify both the between-mode and within-mode correlations in industrial processes, resulting in superior partitioning performance.

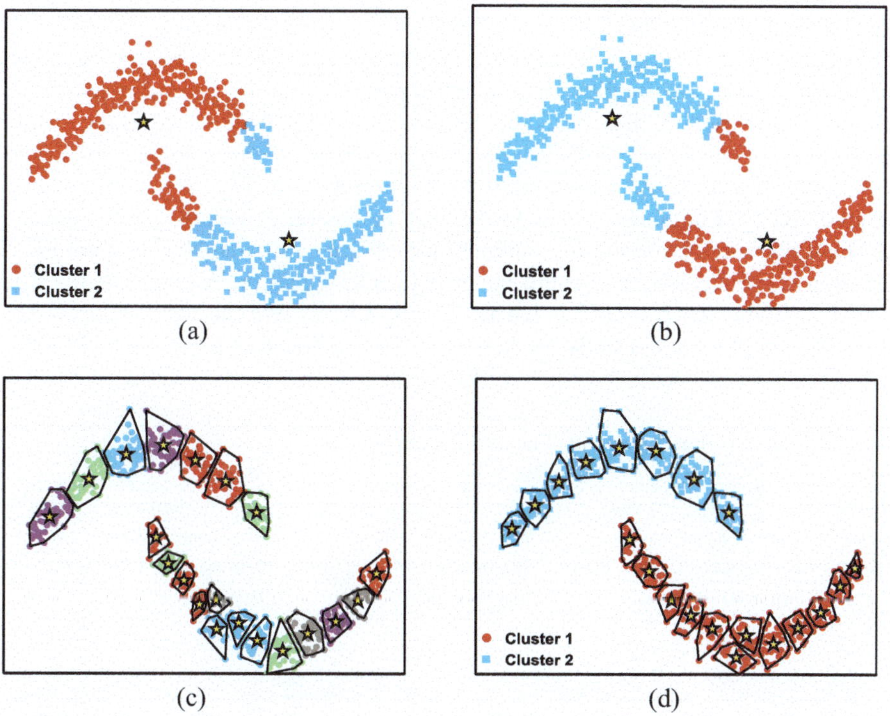

Fig. 7.7 The clustering results of different algorithms on the Two Moons datasets. **a** K-Means ($K = 2$); **b** Fuzzy C-Means ($C = 2$); **c** K-Means ($K = 20$); **d** Robust K-Multiple-Means ($K = 2, M = 20$)

7.3.1 The Numerical Example

Given a numerical example with five measured variables as follow:

$$\mathbf{X} = \begin{bmatrix} x_1 \\ x_2 \\ x_3 \\ x_4 \\ x_5 \end{bmatrix} = \begin{bmatrix} 1 & 0 \\ 0 & 1 \\ 1.57 & 3 \\ 2.7 & 3 \\ 1.2 & 1.6 \end{bmatrix} \begin{bmatrix} s_1 & s_2 \end{bmatrix} + \begin{bmatrix} 0.2e \\ 0.1e \\ 0.2e \\ 0.1e \\ 0.1e \end{bmatrix} \tag{7.19}$$

where $e \sim N(0, 1)$ is white Gaussian noise. Within this numerical example, three distinct states of s_i are specified to mimic three typical modes of the process. Then, faults are then introduced into each of these modes. As a result, each mode encompasses four types of samples: one representing a normal condition and three representing faulty conditions. The comprehensive fault information is outlined in Table 7.2. The goal of this numerical example is to emulate the varied characteristics observed in multi-mode fault processes. In this chapter, we generate 200 samples for each type within each mode, resulting in a total of 2400 samples. Half of these samples (50%) are used as the training data, while the remaining half (50%) are used as the testing data.

Initially, results of the mode partition are presented in Fig. 7.8a. The method based on RKMM demonstrates its proficiency in accurately identifying the modes, highlighting its effectiveness in mode partition. Figure 7.8b displays the fault identification outcomes obtained using our framework, where the normal condition is labeled as 0 and the faults are labeled as 1-3, respectively. We can see that the classes of testing samples are correctly predicted, with a classification accuracy of 98.83%.

Table 7.2 The experimental setting of numerical example

Status	Mode 1	Mode 2	Mode 3
Normal	$s_1 \sim N(0, 1)$, $s_2 \sim N(0, 1)$	$s_1 \sim N(10, 0.8)$, $s_2 \sim N(12, 1.3)$	$s_1 \sim N(-10, 0.5)$, $s_2 \sim N(0.5, 3)$
Fault 1	Step change of 2.5 to s_1	Step change of 2.5 to s_1	Step change of 3.5 to s_1
Fault 2	Step change of 2.5 to s_2, 1 to x_3	Step change of 0.5 to s_2, 1.5 to x_3	Step change of 1.5 to s_2, 1 to x_3
Fault 3	Step change of 2 to x_4	Step change of 2 to x_4	Step change of 2 to x_4

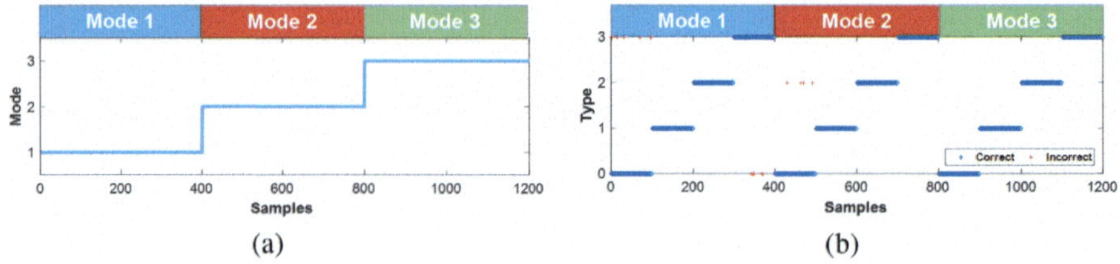

Fig. 7.8 Experimental results in the numerical experiment. **a** Mode identification; **b** Fault identification

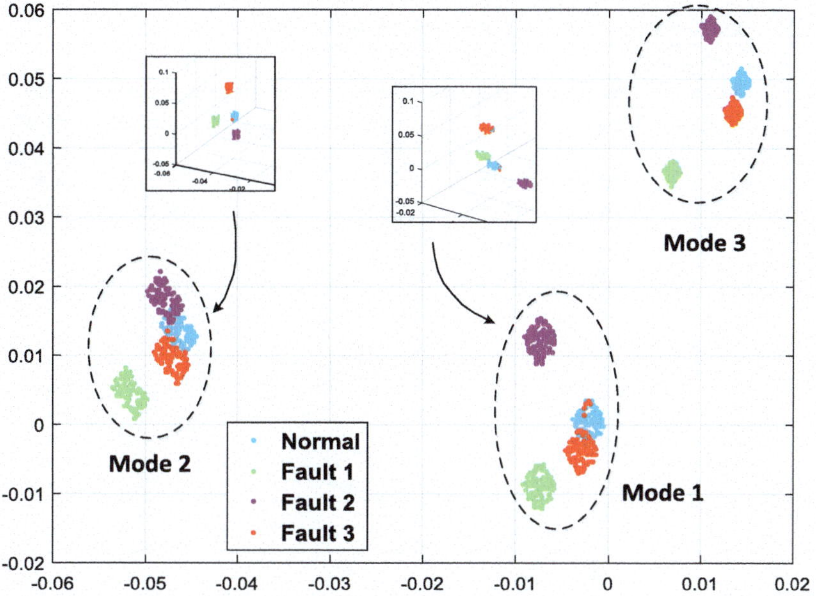

Fig. 7.9 The 2D plot of testing samples in the numerical experiment

To provide a clearer depiction of the numerical example, we offer a 2-D scatter plot depicting the 1200 testing samples. In Fig. 7.9, it is apparent that faults have arisen across multiple modes. It is difficult to identify three simple straightforward hyperplanes to segment these samples into four classes. However, by applying our framework, we successfully partition all samples into three modes, with each mode accurately encompassing the relevant faults. Once the multi-grained structure is identified, the multi-mode problem is transformed into a series of linear classification tasks involving Gaussian distributions. Consequently, impressive diagnostic performance is attained.

7.3.2 Multi-phase Flow Process

In this section, we evaluate the proposed method by considering four fault cases and a normal case in Cranfield Multi-phase Flow Process (CMFP). Table 7.3 summarizes the acquisition of 8000 samples from four operational conditions, simulated by altering two input variables. Specifically, for each class listed in Table 7.3, 50% of the samples are used as the training data, while the remaining 50% are used as the testing data. Note that the process data are inherently non-Gaussian due to the multiple operating states. Figure 7.10 visually assesses the distribution of the 1st, 10th and 23rd observations in the process. This figure plots each data point in x_1, x_{10}

Table 7.3 Details of the considered data in multi-phase flow process

	Mode info		Fault info				
	Water flow rate (kg/s)	Air flow rate (m3/s)	Normal #1	Air line blockage #2	Water line blockage #3	Separator blockage #4	Open direct bypass #5
Mode 1	2	0.0417	1000	500	–	–	500
Mode 2	3.5	0.0208	1500	500	–	500	500
Mode 3	2	0.0278	1000	–	500	500	–
Mode 4	3.5	0.0417	500	–	500	–	–

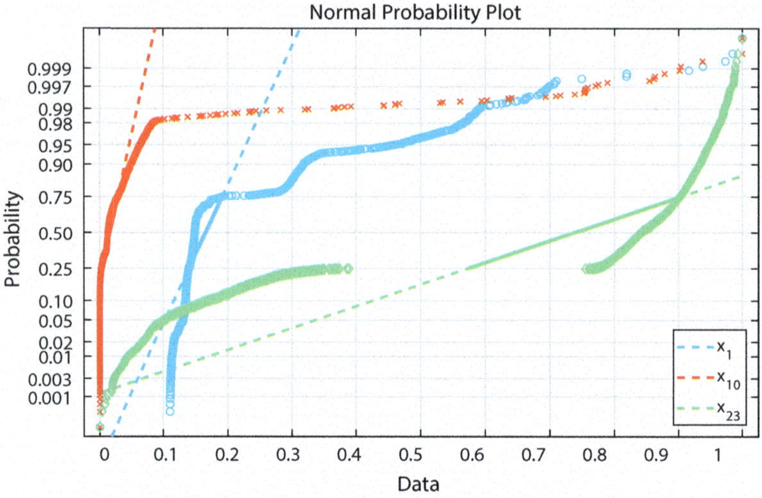

Fig. 7.10 Gaussian probability of the 1st, 10th, 23rd observations in MFP

and x_{23}, with the ordinate representing the cumulative probability and the abscissa representing the data value. If the sample data follows a normal distribution, the data points would align along the reference line. However, it is evident that these data do not adhere to a normal distribution, as they introduce curvature in the plot and deviate from the reference line. This non-Gaussian distribution adds complexity to the fault identification task.

Our primary focus is on fault detection tasks, which can be framed as a binary classification problem. Specifically, class #1 is assigned a negative label, while classes #2-5 are assigned positive labels. The methodologies we compare against include multivariate statistical analysis methods and those based on NMF based, such as PCA, KPCA, Linear Discriminant Analysis (LDA), NMF, Robust NMF [15] and Kernel NMF (with polynomial kernel PNMF and Gaussian kernel GNMF). Notably, PCA, LDA, NMF and RNMF are linear methods. In contrast, KPCA, PNMF and GNMF are nonlinear methods that utilize polynomial or Gaussian kernel functions

Table 7.4 Fault detection performance of different methods on MFP

Method	Miss alarm rate (%)	False alarm rate (%)
PCA	15.60	23.72
LDA	19.65	20.37
KPCA	11.75	14.36
NMF	24.80	23.23
RNMF	19.30	10.98
GNMF	23.55	11.67
PNMF	16.85	15.84
Ours	**11.75**	**8.64**

to map data into a higher-dimensional space. For our proposed framework, we set the two crucial parameters K and M to 4 and 600, respectively, and p to 0.5. The performance of fault detection is evaluated using Miss Alarm Rate (MAR) and False Alarm Rate (FAR),[2] which are calculated as follows:

$$
\begin{aligned}
MAR &= \frac{FN}{TP + FN} \\
FAR &= \frac{FP}{TP + FP}
\end{aligned}
\tag{7.20}
$$

The results of the fault detection are presented in Table 7.4. Upon examination, it is evident that the proposed model outperforms all other competitors in terms of both the MAR and FAR.

Next, we present the comparative results for the fault identification task. The experimental conditions mirror those of the preceding experiment. To assess performance, we utilize the true positive rate (TPR) and positive predictive value (PPV), defined as follows:

$$
\begin{aligned}
TPR &= n_i^t/n_i \times 100\%, \quad Avg(TPR) = \sum_{i=1}^{C} \times TPR(i)/n \\
PPV &= n_i^t/n_i^p \times 100\%, \quad Avg(PPV) = \sum_{i=1}^{C} \times PPV(i)/n
\end{aligned}
\tag{7.21}
$$

where C represents the total number of classes. n is the number of samples. n_i^t, n_i denote the number of correctly diagnosed samples and the total number of samples, respectively, for the i-th fault type. n_i^p refers to the number of samples predicted as the i-th fault. The classification results delivered by our framework as well as competitors are listed in Tables 7.5 and 7.6. Noteworthy observations include the

[2] FN-False Negative. FP-False Positive. TP-True Positive.

Table 7.5 Fault identification performance of different methods on MFP in terms of TPR/

	PCA (%)	LDA (%)	KPCA (%)	NMF (%)	RNMF (%)	GNMF (%)	PNMF (%)	Ours (%)
#1	73.75	79.45	85.20	77.25	90.05	89.90	84.35	91.80
#2	50.00	72.40	65.40	24.80	34.20	17.60	52.20	64.00
#3	100.00	82.20	92.80	100.00	92.40	88.80	90.00	98.60
#4	88.40	20.80	86.80	71.20	76.80	88.20	90.00	76.00
#5	42.40	97.40	64.20	84.60	98.00	93.20	100.00	100.00
AVG	71.98	73.83	81.25	73.70	82.70	80.93	83.70	88.23

Table 7.6 Fault identification performance of different methods on MFP in terms of PPV

	PCA (%)	LDA (%)	KPCA (%)	NMF (%)	RNMF (%)	GNMF (%)	PNMF (%)	Ours (%)
#1	73.75	79.45	87.88	77.25	90.05	89.90	84.35	88.57
#2	50.00	72.40	64.12	24.80	34.20	17.60	52.20	68.52
#3	100.00	82.20	84.21	100.00	92.40	88.80	90.00	98.60
#4	88.40	20.80	68.03	71.20	76.80	88.20	90.00	97.44
#5	42.40	97.40	88.67	84.60	98.00	93.20	100.00	87.72
AVG	71.98	73.83	78.79	73.70	82.70	80.93	83.70	88.32

following:

- When compared to standard NMF, the extended versions of NMF demonstrate superior performance in terms of both TPR and PPV. This suggests that appropriate modifications can enhance the diagnostic capabilities of NMF.
- In practical applications, the parameters for polynomial and Gaussian kernel functions are typically determined empirically. Through trial and error, it has been observed that incorrect parameter settings can lead to a decrease in classification accuracy. Consequently, the performance of kernel-based nonlinear methods is highly dependent on the choice of parameters.
- The average TPR and PPV of our proposed framework are 88.23% and 88.32%, respectively, representing an improvement of 14.54% and 15.25% over those of the conventional NMF. Our framework also outperforms other comparable methods, achieving an additional increase of 4.53% in TPR and 4.62% in PPV compared to the best competitor. These results indicate the effectiveness of our framework.

The robustness of the proposed framework is demonstrated through a discussion on optimal norm selection. The parameter p varied within the set $\{0.15, 0.25, 0.5, 0.75, 1, 1.5, 2\}$ across different noise amplitudes ranging from 10% to 50%. The identification performance for each p value is illustrated in Fig. 7.11. The ordinates denote the metrics in terms of PPV and TPR, and the abscissas highlight p values that were chosen. Analysis of Fig. 7.11 reveals that when the noise

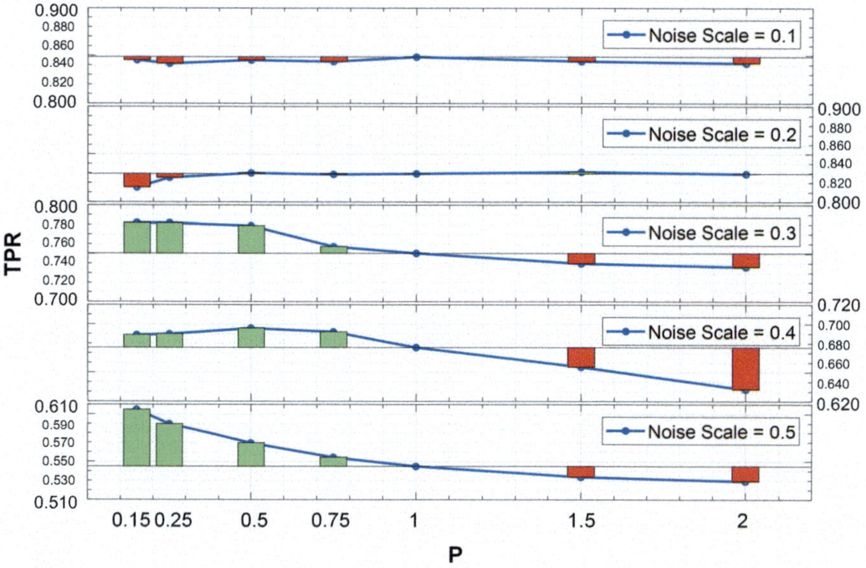

(a)

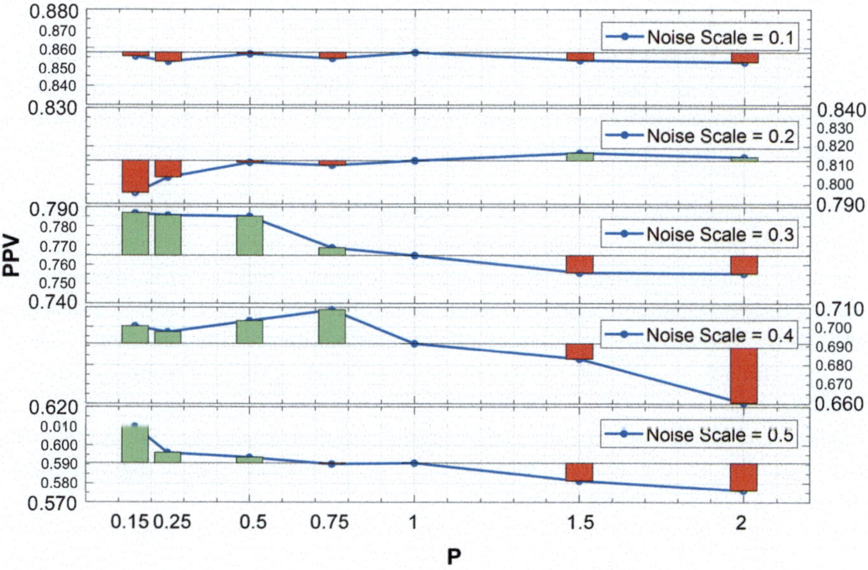

(b)

Fig. 7.11 Identification performance under different noise scales when p varies. **a** performance in terms of TPR, **b** performance in terms of PPV

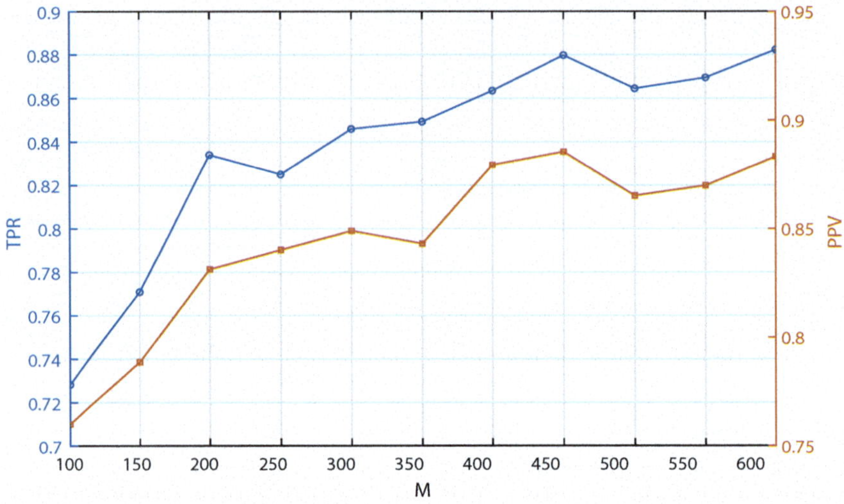

Fig. 7.12 Identification performance when M varies

amplitude is low (noise scale = 0.1/0.2), variations in p have a negligible effect on diagnostic performance. However, as the noise amplitude increases (noise scale $\geq$ 0.3), cases where p is smaller than 1 prevail over cases where p is larger than 1. This fact demonstrates that the $\ell_p(p < 1)$ is able to handle the noise in MFP and achieve better identification performance, both on TPR and PPV. Therefore, we recommend setting the parameter p within the range of [0.25, 1] to obtain satisfactory diagnostic results.

Regarding another crucial parameter, denoted as M, Fig. 7.12 depicts the variations in TPR and PPV as the parameter M is varied. As we can see, a smaller value of M fails to attain satisfactory diagnostic performance because it is inadequate in ensuring the precise identification of detailed information. Conversely, increasing M significantly leads to substantial demands on computational time and space. Therefore, selecting $M = \sqrt{N \times K}$ (200 in this experiment) might be a compromising but promising way.

The convergence curve of the RNMF is provided in Fig. 7.13. We can see that the proposed method converges very fast within 20 iterations, showing agreements of the theoretical convergence analysis.

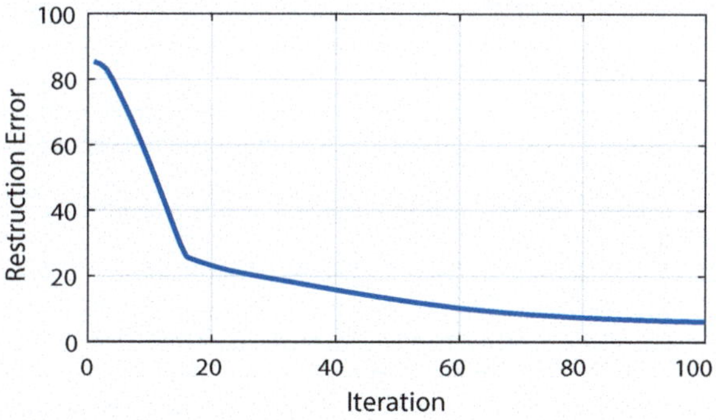

Fig. 7.13 The convergence curve of the RNMF

7.4 Conclusion

This chapter introduces a novel framework for fault identification in multi-mode industrial processes, structured into three components: mode partition, feature learning and fault identification. Specifically, the RKMM-based clustering algorithm leverages multi-grained information to discover non-Gaussian (even non-convex) mode distributions, enabling precise mode partition. The proposed RNMF method ensures that each sample is accurately represented by a nonnegative combination of the discovered multi-grained partition information. Then, the KNN classifier with the weighted metric, effectively measures fault similarity for classification purposes. Moreover, the ℓ_p-norm within our framework robustly addresses noise and outliers in industrial settings. The proposed framework is validated on a numerical example and a real industrial process. Experimental results indicate that our framework achieves satisfactory diagnostic performance.

References

1. Huang K, Wei K, Zhou L, Li Y, Yang C (2021) Multimode process monitoring and mode identification based on multiple dictionary learning. IEEE Trans Instrum Measur 70:1–12
2. Wang B, Li Z, Dai Z, Lawrence N, Yan X (2019) Data-driven mode identification and unsupervised fault detection for nonlinear multimode processes. IEEE Trans Indus Inform 16(6):3651–3661
3. Peng X, Tang Y, Wenli D, Qian F (2017) Multimode process monitoring and fault detection: A sparse modeling and dictionary learning method. IEEE Trans Indus Electron 64(6):4866–4875
4. Tong C, Palazoglu A, Yan X (2013) An adaptive multimode process monitoring strategy based on mode clustering and mode unfolding. J Process Control 23(10):1497–1507
5. Wanke Yu, Zhao C (2018) Online fault diagnosis in industrial processes using multimodel exponential discriminant analysis algorithm. IEEE Trans Control Syst Technol 27(3):1317–1325
6. Song B, Yan H, Shi H, Tan S (2020) Multisubspace elastic network for multimode quality-related process monitoring. IEEE Trans Indus Inform 16(9):5874–5883
7. Xue X, Ding J, Liu Q, Chai T (2021) A novel multimanifold joint projections model for multimode process monitoring. IEEE Trans Indus Inform 17(9):5961–5970

8. Ma L, Dong J, Peng K (2018) Root cause diagnosis of quality-related faults in industrial multimode processes using robust gaussian mixture model and transfer entropy. Neurocomputing 285:60–73
9. Shao W, Ge Z, Yao L, Song Z (2019) Bayesian nonlinear gaussian mixture regression and its application to virtual sensing for multimode industrial processes. IEEE Trans Autom Sci Eng 17(2):871–885
10. Chung FRK, Graham FC (1997) Spectral graph theory. Number 92. American Mathematical Soc
11. Nie F, Wang C-L, Li X (2019) K-multiple-means: a multiple-means clustering method with specified k clusters. In: Proceedings of the 25th ACM SIGKDD international conference on knowledge discovery and data mining, pp 959–967
12. Chen Z, LiuC, Ding SX, Peng T, Yang C, Gui W, Shardt YAW (2020) A just-in-time-learning-aided canonical correlation analysis method for multimode process monitoring and fault detection. IEEE Trans Indus Electron 68(6):5259–5270
13. Peng K, Zhang K, You B, Dong J, Wang Z (2016) A quality-based nonlinear fault diagnosis framework focusing on industrial multimode batch processes. IEEE Trans Indus Electron 63(4):2615–2624
14. Jie Yu (2013) A new fault diagnosis method of multimode processes using bayesian inference based gaussian mixture contribution decomposition. Eng Appl Artif Intell 26(1):456–466
15. Guan N, Tao D, Luo Z, Yuan B (2012) Online nonnegative matrix factorization with robust stochastic approximation. IEEE Trans Neural Networks Learning Syst 23(7):1087–1099

Chapter 8
Knowledge Dictionary Transferring for Processes Fault Identification with Zero Data Sample

Zero-sample fault identification (ZSFD) has gained significant attention due to its notable success. However, current methodologies face challenges related to fault attribute labeling and feature extraction, which hinder their generalization and robustness. Specifically, relying on expert knowledge for fault attribute labeling is both time-intensive and laborious. Additionally, existing approaches typically extract fault features within a single projection space, disregarding valuable information from target data, thereby leading to suboptimal ZSFD performance. To overcome these challenges, we introduce an innovative zero-sample fault identification method based on transfer learning. Our approach begins by transferring a shared knowledge dictionary, automatically learned from labeled source data, to the target data, thereby reducing ZSFD's reliance on detailed fault descriptions. Next, we introduce a novel multiclass space projection model to capture discriminative fault features. Finally, we employ a pseudolabel mechanism to uncover inter-class and intra-class information within the target domain. Experiments with the Tennessee-Eastman process and a real-world hot roll steel process showcase the effectiveness and superiority of the method.

8.1 Motivation

In reality, fault samples are often scarce and many types of fault are destructive, limiting the availability of sample data [1]. Furthermore, it is challenging to collect fault data in complex systems, as it is not feasible to allow the equipment to run to the point of failure [2]. The long operational periods required for complex equipment to transition from normal to faulty states further complicate data collection [2]. As a result, ZSFD has become a vital research focus within industrial predictive maintenance technology. Zero-shot learning, initially utilized for animal recognition tasks [3], focuses on training models to classify animals from unseen classes (target domain) by using knowledge from known classes (source domain) with the help

H. Yin et al., *Data-Driven Fault Diagnosis for Complex Industrial Processes*, Engineering Applications of Computational Methods 22,
https://doi.org/10.1007/978-981-96-3153-7_8

of semantic attributes. Lampert et al. [3] introduced the direct attribute prediction (DAP) method, which involved learning animal attributes like "color" and "striped" from tiger and panda images to classify zebras and bees. They later expanded this approach with indirect attribute prediction (IAP), which learns animal attributes from labels [4]. Zero-shot learning can also leverage implicit knowledge from the source domain, as demonstrated by Mensink et al. [5], who performed zero-shot image classification by learning visual concept cooccurrences in images. Recently, Feng et al. [6] applied ZSFD to industrial processes, and Xu et al. [7] explored zero-shot learning for compound fault identification in bearings. Existing zero-shot fault identification approaches generally fall into attribute space-based zero-shot learning [1, 8] and domain adaptation-based zero-shot learning [9, 10].

Attribute space-based methods aim to develop a set of attributes from a large number of labeled samples and perform label projection. These attributes are then used to learn the representation of target faults, which can be classified using various metrics or classifiers (as depicted in Fig. 8.1a). Xing et al. [8] proposed a label description space embedding model for fault identification, which constructs descriptions for compound faults by utilizing a set of individual faults within an attribute space. Xu et al. [11] created a semantic feature encoding method to characterize unknown faults by establishing an attribute space based on the vibration features of original signals. Similarly, Chen et al. [12] developed a deep attention relation network to extract features from a known domain and apply them to an unknown domain. While these methods have been designed specifically for bearing fault identification, their effectiveness for more complex industrial processes remains unexplored. Additionally, Feng et al. [6] utilized human-defined fault descriptions instead of individual fault samples, learning fault attributes based on these handcrafted descriptions. Zhou et al. [1] took a different approach by generating rare samples from the same attribute in source data using a generative adversarial network (GAN), and then training the classifier with these generated samples. Despite this difference, the attributes of the rare samples are still annotated by skilled experts.

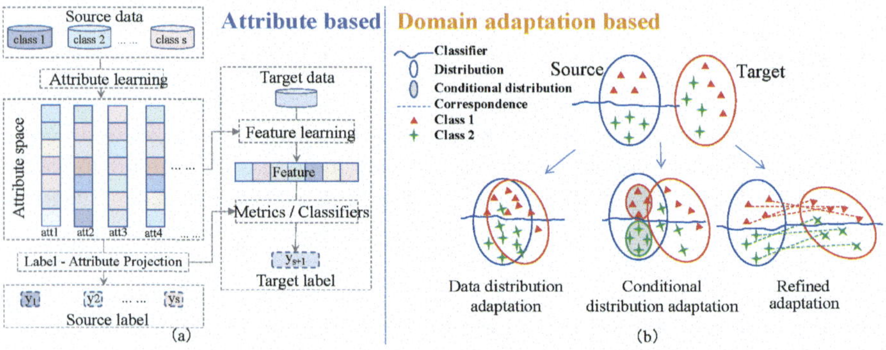

Fig. 8.1 Two kinds of the existing work related to zero sample fault identification

Domain adaptation methods are particularly effective for addressing cross-domain fault identification rather than zero-sample fault identification (ZSFD). These methods work by aligning the distribution between source and target data to mitigate domain bias (as illustrated in Fig. 8.1b). For example, Jiao et al. [13] introduced a residual joint adaptation adversarial network to correct data distribution discrepancies across different working conditions. Sun et al. [14] developed a stacked auto-encoder algorithm for transfer learning to balance different categories. Qin et al. [15] developed a deep joint adaptation mechanism designed to align marginal distributions across source and target domains through an innovative statistical metric. To further address domain bias, researchers have developed sub-domain adaptation methods and other refined techniques. Wang et al. [9] identified the limitations of global domain alignment and proposed a sub-domain adaptation transfer learning network to adapt both marginal and conditional distribution biases. Li et al. [16] created an attention-based deep meta-transfer learning method for few-shot fine-grained fault identification. Wang et al. [17] used an attention mechanism to align local distributions between different domains. Chai et al. [10] introduced a multi-source refined transfer network aimed at enhancing adversarial adaptations from source to target data.

The current methods face three major limitations, leading to sub-optimal generalization and accuracy in ZSFD algorithms.

- Existing methods rely extensively on expert knowledge to obtain detailed data descriptions and attributes. However, such prior information is not always available for all faults. As a result, there is no assurance that these methods will perform effectively for new faults without any prior information.
- While current domain adaptation techniques have evolved from global transfer to local refined transfer to mitigate negative effects, suboptimal features are still often extracted due to the fault data being projected into the same space. This results in less accurate fault identification.
- Existing methods primarily focus on exploring intra-class and inter-class information within the source domain due to the abundance of labeled data, while the target domain lacks such label information. However, this information can be explored in a semi-supervised manner within the target domain and used to enhance fault identification accuracy.

To overcome these limitations, this chapter presents a novel ZSFD approach founded on transfer learning, aiming to improve ZSFD accuracy by identifying optimal fault features and valuable target data information without relying on fault description and attribution. Firstly, inspired by dictionary learning [18], a shared knowledge structure is created to reduce ZSFD's dependence on fault descriptions. This structure stores essential knowledge from training data in a dictionary, facilitating feature learning for new data through dictionary lookups. Secondly, a multi-space projection model is proposed to extract optimal features, with each category assigned different projection spaces before feature learning. Lastly, a pseudo-label mechanism is implemented to determine the pseudo-labels of target data. This mechanism utilizes the relationships between source and target data to explore intra-class and inter-class information within the target domain. The main contributions are as follows:

- The proposed shared knowledge dictionary structure eliminates the need for extensive expertise to establish fault attributes, making it more widely applicable to zero-sample diagnostics across various fault types.
- A new multi-space strategy is introduced to extract optimal features by using projection subspaces derived from various categories.
- To leverage the valuable information within target data, this chapter presents a pseudo-label mechanism designed to uncover the intra-class and inter-class characteristics within the target domain.

8.2 Problem Formulation

Suppose $X = [x_1, x_2, \ldots, x_N]^\top$ represents the fault samples, and $Y = [y_1, y_2, \ldots, y_N]^\top$ represents the fault class labels, where N is the overall count of fault samples. Let x_i denote the i-th fault sample, and y_i be the class label of the i-th fault sample. The goal of fault identification is to identify determine the fault class label y_{test} for a new fault sample x_{test} using a trained diagnostic model. This can be formulated as follows:

$$Y_{test} = f(X_{test}|X_{train}, Y_{train}), \tag{8.1}$$

here, f denotes the trained identification model. In the typical case, the labels of training samples and testing samples belong to the same set of fault classes ($Y_{train} \cup Y_{test} = Y_{train}$ or Y_{test}). However, it is common to encounter situations where no labeled fault samples are available for training, meaning $Y_{train} \cap Y_{test} = \varnothing$. This scenario is known as ZSFD.

Zero-shot learning offers a potential solution for ZSFD by utilizing other relevant but unlabeled fault samples to address the problem. Some notations are introduced to elaborate on ZSFD. Following the research of Sinno Jialin Pan et al. [19], a domain is denoted as $\mathcal{D} = \{X, P(X)\}$, where $P(X)$ represents the marginal probability distribution. Consequently, the source domain and target domain are denoted as $\mathcal{D}_S = \{X_S, P(X_S)\}$ and $\mathcal{D}_T = \{X_T, P(X_T)\}$, where X_S and X_T represent the features of source samples and target samples, respectively. The ZSFD problem can be formulated as:

$$\begin{aligned} Y_T &= f_{ZSFD}\left\{f_{\mathcal{D}_T}|(f_{\mathcal{D}_S}, X_T)\right\} \\ s.t. X_S &\cap X_T = \varnothing, Y_S \cap Y_T = \varnothing \\ f_{\mathcal{D}_S} &: X_S \rightarrow Y_S, f_{\mathcal{D}_T} : X_T \rightarrow Y_T, \end{aligned} \tag{8.2}$$

where $f_{\mathcal{D}_S}$ is the objective function for the source domain, $f_{\mathcal{D}_T}$ is the objective function for the target domain, and F_{ZSFD} is the objective function for ZSFD. Here, Y_S and Y_T represent the labels of the source and target samples, respectively.

8.3 Methodology

In this section, we outline the newly proposed semi-supervised ZSFD method based on transfer learning. Figure 8.2 illustrates the complete framework of our approach. A unique projection space is assigned to each category to mitigate the negative effects on each class and prevent sample bias during the projection process. By assigning one class to one space, we can fully exploit the distinguishing features hidden within the data. Next, a set of fundamental bases, representing shared knowledge in the source domains, is extracted to aid in the optimal projection of target data. Additionally, the k-means clustering approach is employed to generate pseudo-labels for target data, capturing inter-class and intra-class information of the unlabeled samples and enhancing classification accuracy.

8.3.1 *Multi-class Space Projection*

In traditional feature extraction, all categories share a single projection space. This often leads to samples from different classes compromising due to the limitations of

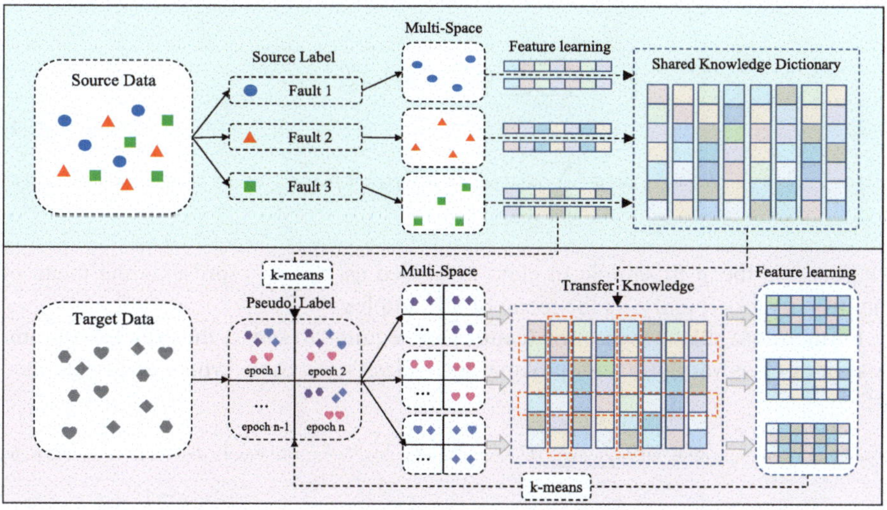

Fig. 8.2 The framework of proposed ZSFD method based on transfer learning

using only one projection space. To address this issue and enhance the discriminate power of projection features, we propose constructing a unique domain for each category using the multi-class space projection method. Let $X_S^i \in \mathbb{R}^{n\times d}$ represent the i-th class of source samples, where n is the total fault sample count, d indicates the feature dimension of each sample, and $i \in \{1, 2, \ldots, s\}$, with s being the count of classes in the source domain. The projection matrix for X_S^i is denoted as $M_S^i \in \mathbb{R}^{d\times m}$, where m is the reduced dimensionality. The projection process can be specified as $\tilde{X}_S^i = X_S^i M_S^i$, where $\tilde{X}_S^i \in \mathbb{R}^{n\times m}$ represents the fault features extracted from X_S^i using the projection matrix M_S^i. According to the properties of Euclidean space, the objective function for the proposed projection matrix M_S^i is formulated as follows:

$$\arg\min_{M_S^i} \sum_{i=1}^{s} \left\| \tilde{X}_S^i - X_S^i M_S^i \right\|_F^2 . \tag{8.3}$$

In traditional feature extraction, all categories utilize a single projection space, often causing samples from different classes to overlap and compromise due to this limitation. To overcome this issue and enhance the discriminative power of projection features, we propose a multi-class space projection method that creates a unique domain for each category. Let $X_S^i \in \mathbb{R}^{n\times d}$ denote the i-th class of source samples, where n is the total count of fault samples, d is each sample's dimensionality, and $i \in \{1, 2, \ldots, s\}$, with s representing the number of classes in the source domain. The projection matrix for X_S^i is denoted as $M_S^i \in \mathbb{R}^{d\times m}$, where m is the reduced dimension. The projection process is designated as $\tilde{X}_S^i = X_S^i M_S^i$, where $\tilde{X}_S^i \in \mathbb{R}^{n\times m}$ represents the fault features extracted from X_S^i using the projection matrix M_S^i. In accordance with the properties of Euclidean space, the function to be optimized for the proposed projection matrix M_S^i is formulated as follows:

$$\begin{aligned} S_w &= \sum_{i=1}^{s} \sum_{y_S^{ip}} (x_S^{ip} - \mu_i)(x_S^{ip} - \mu_i)^\top \\ S_b &= \sum_{i=1}^{s} n_i (\mu_i - \mu)(\mu_i - \mu)^\top , \end{aligned} \tag{8.4}$$

where x_S^{ip} is the p-th sample in class i, labeled as y_S^{ip}. μ_i represents the mean of samples in class i, and μ is the mean of all samples.

Furthermore, all the projection features are required to satisfy the principle of minimal intra-class variation and maximal inter-class separation, which can be defined as:

$$\arg\min_{\tilde{x}_S^{ip}} \sum_{i=1}^{s} \sum_{y_S^{ip}} tr(\tilde{x}_S^{ip} S_w (\tilde{x}_S^{ip})^\top - \tilde{x}_S^{ip} S_b (\tilde{x}_S^{ip})^\top), \tag{8.5}$$

where, $\tilde{x_S^{ip}} S_w (\tilde{x_S^{ip}})^\top$ and $\tilde{x_S^{ip}} S_b (\tilde{x_S^{ip}})^\top$ indicate the intra-class scatter and the inter-class scatter, respectively.

Finally, the multi-class space projection objective function, incorporating prior label information, can be expressed as:

$$\arg\min_{M_S^i, \tilde{X_S^i}} \sum_{i=1}^{s} \left\{ \begin{array}{c} \alpha_s \left\| \tilde{X_S^i} - X_S^i M_S^i \right\|_F^2 + \left\| M_S^i \right\|_F^2 \\ tr(\tilde{X_S^i} S_w (\tilde{X_S^i})^\top - \tilde{X_S^i} S_b (\tilde{X_S^i})^\top) \end{array} \right\}, \tag{8.6}$$

where α_s is a parameter and $\| M_S^i \|_F^2$ serves as a regularization term to prevent overfitting. Equation (8.6) aims to create a robust projection space for each category, resulting in projected features with improved separability. The term $\alpha_s \left\| \tilde{X_S^i} - X_S^i M_S^i \right\|_F^2 + \| M_S^i \|_F^2$ represents the multi-class space projection model.

In summary, the primary objective of the source domain is to derive a set of projection matrices M_S using the labeled information. This approach minimizes the distance between samples within the same class, while maximizing the distance between samples of different classes.

The multi-class space projection model for the target domain is defined in a similar manner as follows:

$$\arg\min_{M_T^j, \tilde{X_T^j}} \sum_{j=1}^{t} \left\{ \begin{array}{c} \alpha_t \left\| \tilde{X_T^j} - X_T^j M_T^j \right\|_F^2 + \left\| M_T^j \right\|_F^2 \\ tr(\tilde{X_T^j} S_w (\tilde{X_T^j})^\top - \tilde{X_T^j} S_b (\tilde{X_T^j})^\top) \end{array} \right\}, \tag{8.7}$$

where $X_T^j \in \mathbb{R}^{k \times h}$ represents the target samples of the j-th class, and $M_T^j \in \mathbb{R}^{h \times l}$ is the projection matrix for X_T^j. The parameter α_t is also introduced.

However, due to the lack of prior knowledge, such as label information, in the target domain, the optimization of Eq. (8.7) requires additional information from the source domain.

8.3.2 Shared Knowledge Transfer

Based on the concept of zero-shot learning, related but non-overlapping source domain samples can be leveraged to assist in solving target tasks. In the context of ZSFD, what type of knowledge or information be transferred starting in the source domain and reaching the target domain? Inspired by the principle of dictionary learning, which aims to construct a basic dictionary containing fundamental knowledge, we can apply a similar approach here. When a new task arises, we only need to extract the relevant fundamental knowledge from this dictionary. Specifically, a basic dictionary can be derived from the original data along with its coefficients. Different data can then be referenced from the dictionary using different coefficients, indicating that the essential knowledge of the original data is encapsulated within the learned dictionary. Consequently, We posit that the knowledge within source domain data can be transferred to the target domain through the knowledge dictionary as an intermediary.

Let $D \in \mathbb{R}^{n \times k}$ represent the knowledge dictionary learned from the source domain, and $G_S \in \mathbb{R}^{k \times m}$ be the sparse coefficient for the source domain features $\tilde{X}_S$. Following the concept of dictionary learning, the construction of the knowledge dictionary D using source domain features $\tilde{X}_S$ be described as:

$$\arg\min_{D} \left\|\tilde{X}_S - DG_S\right\|_F^2 + \left\|D\right\|_F^2 + \left\|G_S\right\|_F^2, \tag{8.8}$$

where the shared knowledge dictionary is constructed using the term $\left\|\tilde{X}_S - DG_S\right\|_F^2$. The terms $\left\|D\right\|_F^2$ and $\left\|G_S\right\|_F^2$ serve as regularization terms for the shared knowledge dictionary and sparse coefficients, respectively. This formulation aims to extract shared knowledge from the fault samples in the source domain. Therefore, the source domain objective function can be modified to:

$$\arg\min_{M_S^i, \tilde{X}_S^i, D, G_S^i} \sum_{i=1}^{s} \left\{ \begin{array}{c} \alpha_s \left\|\tilde{X}_S^i - X_S^i M_S^i\right\|_F^2 + \left\|M_S^i\right\|_F^2 + \\ tr(\tilde{X}_S^i S_w (\tilde{X}_S^i)^\top - \tilde{X}_S^i S_b (\tilde{X}_S^i)^\top) \\ +\beta_s \left\|\tilde{X}_S^i - DG_S^i\right\|_F^2 + \left\|D\right\|_F^2 + \left\|G_S\right\|_F^2 \end{array} \right\}, \tag{8.9}$$

where β_s is a parameter intended to balance the update speed of the objective function. After learning the knowledge dictionary from the source domain, we need to obtain the target domain features $\tilde{X}_T$ by referencing the knowledge dictionary. This can be expressed as:

$$\arg\min_{M_T^j, \tilde{X}_T^j, G_T^j} \sum_{j=1}^{t} \left\{ \begin{array}{c} \alpha_t \left\|\tilde{X}_T^j - X_T^j M_T^j\right\|_F^2 + \left\|M_T^j\right\|_F^2 + \\ tr(\tilde{X}_T^j S_w (\tilde{X}_T^j)^\top - \tilde{X}_T^j S_b (\tilde{X}_T^j)^\top) \\ +\beta_t \left\|\tilde{X}_T^j - DG_T^j\right\|_F^2 + \left\|D\right\|_F^2 + \left\|G_T\right\|_F^2 \end{array} \right\}. \tag{8.10}$$

As described in section III.A, the multi-class projection space of the source domain can be determined when label information is available. Therefore, the optimal solution for Eq. (8.9) can be obtained using a general gradient descent algorithm. However, the function in Eq. (8.10) cannot be solved due to the absence of labeled information. Inspired by Lee et al. [20], who proposed that pseudo-labels are effectively equivalent to entropy regularization and reduce the information content of unlabeled examples with class overlap, we designed a pseudo-label mechanism based on the relationship between source and target domain data. Although theoretical support is limited, experimental evidence highlights the benefits of using unlabeled examples in the learning process. In parametric statistics, the use of unlabeled examples has proven advantageous [21]. Thus, incorporating pseudo-labels benefits model training, even when the labels are incorrect, as their impact diminishes over iterations. The relationship matrix across source and target domain data is represented as $H \in \mathbb{R}^{N_s \times N_t}$, where N_s is the number of source samples and N_t is the number of target samples. H_{ij} denotes the distance from source sample i to target sample j. Additionally, the K-means clustering algorithm is chosen to derive pseudo-labels from the correlation matrix H due to its efficiency and relatively low computational complexity. Another advantage of using pseudo-labels is that the inter-class and intra-class information of the target domain, previously neglected, is utilized to improve fault identification accuracy. This process involves iterating clustering and alternating updates between the source and target domains until convergence.

8.3.3 Optimization

This chapter employs an efficient alternating optimization algorithm to decompose the multi-variable optimization problem into several single-variable optimization subproblems. Before optimizing function Eq. (8.10), the knowledge dictionary D must be derived from function Eq. (8.9). It's essential to recognize that function Eq. (8.10) is similar to objective Eq. (8.9), differing only in the domains (source domain S and target domain T) and the number of classes (s and t). Therefore, the optimization approach for the target domain objective is analogous to that of the source domain. Let the number of classes be denoted as $c = 1, 2, \ldots, cnum$ $(cnum = s/t)$. The functions Eqs. (8.9) and (8.10) can be formulated as follows:

$$f_1\{M^c, \tilde{X^c}\} = \sum_{c=1}^{cnum} \left\{\alpha_c \left\|\tilde{X^c} - X^c M^c\right\|_F^2 + \left\|M^c\right\|_F^2\right\}, \tag{8.11}$$

$$f_2\{\tilde{X^c}\} = \sum_{c=1}^{cnum} \left\{tr(\tilde{X^c} S_w (\tilde{X^c})^\top - \tilde{X^c} S_b (\tilde{X^c})^\top)\right\}, \tag{8.12}$$

$$f_3\{D, G^c\} = \sum_{c=1}^{cnum} \beta_c \left\|\tilde{X^c} - DG^c\right\|_F^2 + \left\|D\right\|_F^2 + \left\|G^c\right\|_F^2. \tag{8.13}$$

Therefore, the objective function can be written as:

$$\mathop{\arg\min}_{M^c, \tilde{X^c}, D, G^c} \{f_1\{M^c, \tilde{X^c}\} + f_2\{\tilde{X^c}\} + f_3\{D, G^c\}\}. \tag{8.14}$$

Based on Eq. (8.14), the variables for both the source domain and the target domain include the projection matrix M^c, the projection feature $\tilde{X}^c$, the knowledge dictionary D, and the sparse coefficient G^c.

8.3.3.1 Update Projection Matrix M^c

By fixing all variables except M^c and equating the derivative of Eq. (8.14) with respect toM^c to zero, we obtain:

$$\frac{\partial f_1\{M^c, \tilde{X}^c\}}{\partial M^c} = \sum_{c=1}^{cnum} \{2\alpha_c (X^c)^\top X^c M^c - 2\alpha_c (X^c)^\top \tilde{X}^c + 2M^c\}. \tag{8.15}$$

Therefore, the equation for updating projection matrix M^c is defined by:

$$M^c_{z+1} = M^c_z - \mu_1 \frac{\partial f_1}{\partial M^c}, \tag{8.16}$$

where z represents the z-th iteration, and μ_1 is the update step size parameter.

8.3.3.2 Update Projection Feature $\tilde{X}^c$

Similarly, by fixing all variables except $\tilde{X}^c$ and applying the gradient descent method, the derivative of Eq. (8.14) with respect to $\tilde{X}^c$ is:

$$\frac{\partial f_1\{M^c, \tilde{X}^c\}}{\partial X^c} = \sum_{c=1}^{cnum} \alpha_c\{-2(X^c)M^c + 2(X^c)\}, \tag{8.17}$$

$$\frac{\partial f_2\{\tilde{X}^c\}}{\partial X^c} = \sum_{c=1}^{cnum} \{2\tilde{X}^c(S_w - S_b)\}. \tag{8.18}$$

Therefore, the equation for updating the projection matrix $\tilde{X}^c$ is written as:

$$\tilde{X}^c{}_{z+1} = \tilde{X}^c_z - \mu_2\{\frac{\partial f_1}{\partial \tilde{X}^c} + \frac{\partial f_2}{\partial \tilde{X}^c}\}, \tag{8.19}$$

where μ_2 is the update step size parameter.

Algorithm 1 Algorithm for ZSFD

Input:

1: The source domain samples X_S^i.
2: The label of source domain samples Y_S^i.
3: The target domain samples X_T^j.

Output:

1: The label of target domain samples Y_T^j.

Initialization:

1: Initialize the project matrix M_S^i.
2: Initialize the project matrix M_T^j.
3: Initialize dictionary D and coefficient G_S^i, G_T^j

Procedure:

1: Obtain the rough label of target domain samples by k-means cluster algorithm.
2: **while** not converged or reach at $maxIter$ **do**
3: Fixing $\tilde{X}_S^i$, D and G_S^i, update M_S^i by Eq. (8.16).
4: Obtain the inter-class matrix S_w and the intra-class matrix S_b for source domain samples by Eq. (8.4).
5: Fixing M_S^i, D and G_S^i, update X_S^i by Eq. (8.19).
6: Fixing M_S^i, $\tilde{X}_S^i$ and G_S^i, update D by Eq. (8.21).
7: Fixing M_S^i, $\tilde{X}_S^i$ and D, update G_S^i by Eq. (8.23).
8: Fixing M_T^j and $\tilde{X}_T^j$, update G_T^j by Eq. (8.23).
9: Update label of target domain samples by k-means clustering H.
10: Obtain the inter-class matrix S_w and the intra-class matrix S_b for source domain samples by Eq. (8.4).
11: Fixing $\tilde{X}_T^j$ and G_T^j, update M_T^j by Eq. (8.16).
12: Fixing M_T^j and G_T^j, update X_T^j by Eq. (8.19).
13: **end while**
14: **return** The cluster label of target domain samples Y_T^j.

8.3.3.3 Update Knowledge Dictionary *D*

Similar to the previous section, by keeping M^c, $\tilde{X}^c$, and G^c fixed, the derivative of Eq. (8.14) with respect to D can be written as:

$$\frac{\partial f_3\{D, G^c\}}{\partial D} = \sum_{c=1}^{cnum}\{-2\beta_c(\tilde{X}^c)(G^c)^\top + 2\beta_c DG^c(G^c)^\top + 2D\}. \tag{8.20}$$

The knowledge dictionary D can be updated by the following function:

$$D_{z+1} = D_z - \mu_3\frac{\partial f_3}{\partial D}, \tag{8.21}$$

where μ_3 is the update step size parameter.

8.3.3.4 Update Sparse Coefficient G

Following the previous steps, by fixing all variables except G and establishing the derivative of Eq. (8.14) in terms of G to zero, we obtain:

$$\begin{aligned}\frac{\partial f_3\{D, G^c\}}{\partial G^c} = \sum_{c=1}^{cnum}\{ - 2\beta_c D^\top \tilde{X}^c \\ + 2\beta_c D^\top D(G^c) + 2G^c\}.\end{aligned} \tag{8.22}$$

Thus, the equation for updating the projection matrix G^i is expressed as:

$$G^i_{z+1} = G^c_z - \mu_4 \frac{\partial f_3}{\partial G^c}, \tag{8.23}$$

where μ_4 is the updated step size parameter. The complete procedure of the proposed method is outlined in the pseudo-code shown in Algorithm 1.

8.3.4 *Feasibility Analysis*

The fault samples in the source domain and target domain are denoted as $X_S = [x_{s1}, x_{s2}, \ldots, x_{sn}]$ and $X_T = [x_{t1}, x_{t2}, \ldots, x_{tn}]$, where sn and tn represent the number of fault samples in the source and target domains, respectively. The fault features of the source domain and target domain are represented as $\tilde{X}_S = [\tilde{x}_{s1}, \tilde{x}_{s2}, \ldots, \tilde{x}_{sn}]^\top$ and $\tilde{X}_T = [\tilde{x}_{t1}, \tilde{x}_{t2}, \ldots, \tilde{x}_{tn}]^\top$, respectively. D_S and D_T denote the knowledge dictionaries for the source and target domains, while G_S and G_T are the coding coefficients for $\tilde{X}_S$ and $\tilde{X}_T$ in accordance with the shared knowledge dictionaries D_S and D_T, respectively.

The proposed ZSDF method seeks to transfer valuable knowledge from the source domain to the target domain, operating under the assumption that knowledge is shared between the two domains, i.e., $D_S = D_T$. This assumption allows us to use fault samples from the source domain to aid in fault identification in the target domain. However, it is crucial to analyze the feasibility of sharing knowledge between the domains.

Definition: For any $\tilde{x}_{ti}$, assume there exists $u_i \in \mathbb{R}^{1\times n}$ such that $\tilde{x}_{ti} = u_i \tilde{X}_S$. Similarly, for any g_{ti}, assume there exists $v_i \in \mathbb{R}^{1\times n}$ such that $g_{ti} = v_i G_S$. Here, u_i and v_i represent the relational knowledge of $\tilde{X}_T$ and G_T concerning $\tilde{X}_S$ and G_S, respectively.

The set $U = \{u_1, u_2, \ldots, u_i\}$ captures a form of dependence between $\tilde{X}_T$ and $\tilde{X}_S$, while the set $V = \{v_1, v_2, \ldots, v_i\}$ encodes similar dependencies in the coefficients of G_T and G_S.

Lemma: If for all j, $v_j = u_j$, then the shared knowledge originating in the source domain D_S can be utilized as the knowledge for the target domain D_T. The relationship between the fault samples in the source domain $\tilde{X}_S$ and the coding coefficients G_S can be described as a mapping $\mathbb{D}_S$: $\tilde{X}_S \to G_S$. Similarly, we have $\mathbb{D}_T$: $\tilde{X}_T \to G_T$.

Proof: Let $\tilde{x}_{tj} = \sum_{i=1}^{sn} u_j \tilde{x}_{si}$. Then, $\mathbb{D}_S(\tilde{x}_{tj}) = \mathbb{D}_S(\sum_{i=1}^{sn} u_j \tilde{x}_{si}) = u_j \mathbb{D}_S(\sum_{i=1}^{sn} \tilde{x}_{si}) = u_j G_S = v_j G_S = g_{tj} = \mathbb{D}_T(\tilde{x}_{tj})$.

Therefore, if $v_i = u_i$ for all i, the shared fault knowledge D between the source domain and the target domain can be effectively utilized to conduct fault identification in the target domain.

8.4 Case Study

8.4.1 Tennessee Eastman Process

8.4.1.1 Data Description

We first use TEP dataset for evaluation. ZSFD tasks differ from traditional fault identification because the categories of test data and training data do not overlap. In this chapter, the training data of the TEP dataset (21 faults) was used for experiments and divided into five groups (Table 8.1, Groups F to J). From these, 12 faults were randomly selected as source domain data, and another 3 faults were chosen as target domain data. Each group was repeated 20 times, considering that all experiments were executed using a semi-supervised clustering algorithm. Notably, there is only one existing study on ZSFD, published by Feng et al. [6] in 2021. To ensure a fair comparison, the proposed method was tested on the same five groups used by Feng et al. (as shown in Table 8.1, Groups A to E).

Table 8.1 The details of the redivided ten groups are as follows, where *S* denotes the source domain data, *T* denotes the target domain data, and '-' indicates the unused fault classes

Groups	1	2	3	4	5	6	7	8	9	10	11	12	13	14	15	16	17	18	19	20	21
A	T	S	S	S	S	T	S	S	S	S	S	S	S	T	S	–	–	–	–	–	–
B	S	S	S	T	S	S	T	S	S	T	S	S	S	S	S	–	–	–	–	–	–
C	S	S	S	S	S	S	S	T	S	S	T	T	S	S	S	–	–	–	–	–	–
D	S	T	T	S	T	S	S	S	S	S	S	S	S	S	S	–	–	–	–	–	–
E	S	S	S	S	S	S	S	S	T	S	S	S	T	S	T	–	–	–	–	–	–
F	S	S	S	S	T	–	–	T	S	S	–	S	–	S	–	S	S	S	S	–	T
G	S	S	S	S	–	S	T	S	T	S	S	T	–	–	S	S	S	S	–	–	–
H	S	–	–	–	T	S	S	S	T	S	S	S	S	S	S	T	–	S	S	–	–
I	S	S	–	–	S	S	S	S	S	S	S	–	T	S	–	S	–	S	T	T	–
J	S	–	–	–	S	T	S	–	–	S	S	T	S	S	S	S	T	S	S	S	–

8.4.1.2 Results and Analysis

The results of the ten groups are shown in Table 8.2 , which includes the accuracy of other approaches such as DAP [22], IAP [23], SJE [24], ESZSL [25], Feng (LSVM, RF, NB) [6], SCS (LSVM, EF, NB) [26], Zhou et al. [1], and k-means [27]. It should be emphasized that our method operates as a semi-supervised algorithm, while the other seven methods, except for the baseline k-means, are supervised algorithms. This means our method does not require the prior setting of transfer label information, such as label mapping knowledge. Additionally, existing ZSFD methods rely heavily on abundant and specialized expert knowledge and experience regarding fault types and attributes. This limitation is why Feng et al. [6] focused their approach only on faults 1 to 15, which have more attribute descriptions compared to faults 16 to 21. In contrast, our method is more suitable for common scenarios where fault attribute descriptions are often insufficient in the real world. Furthermore, the k-means method results from direct clustering of the original data, while our method clusters the features derived from models of multi-class space projection and source knowledge transfer using k-means.

From Table 8.2, three key observations can be made:

Firstly, it is evident that useful knowledge can be extracted from different but related faults, significantly improving ZSFD accuracy. The results of our method across all ten groups surpass those of the k-means method, with an average accuracy of 72.79%. Although the accuracy for Group D is lower than that of the SCE method, the accuracy for Groups A, B, C, and E, as well as the overall average, is higher than all the compared methods listed in Table 8.2. Secondly, it is generally known that semi-supervised methods tend to perform worse than supervised methods, as seen with the k-means method having lower accuracy than the method of Zheng et al. However, our method outperforms the other six approaches listed in Table 8.2. This demonstrates that the proposed ZSFD method, which is founded on transfer learning and multi-class space projection, can extract robust and distinctive features. Thirdly, analyzing both Table I and Table II reveals that Groups C (including fault 11), E (including fault 9), and F (including fault 21) have noticeably lower accuracy compared to the other seven groups. These groups present more challenges for fault identification. Despite these difficulties, the proposed method still achieves the best overall results, highlighting its effectiveness.

To illustrate the effectiveness of our method more intuitively, we employ the t-SNE method to visualize both the original fault features and the features extracted by our method. Figure 8.3 displays eight groups of target domain data. It is evident that the fault features in the original space are less discriminative and almost intermixed (Fig. 8.3a, b, e, f). Conversely, the other visualizations in Fig. 8.3 indicate that samples within the same categories are closely clustered, while most samples from different categories are well-separated, even without label information. This demonstrates that the multi-class space projection effectively projects features for both labeled and unlabeled data.

Table 8.2 The result of our method and compared methods on ten groups

Method	Year	A	B	C	D	E	F	G	H	I	J	Average (A-E)	Average (F-J)
DAP [22]	2009	0.5416	0.6263	0.4013	0.5548	0.3694	–	–	–	–	–	0.4986	–
IAP [23]	2014	0.5548	0.6069	0.4500	0.3625	0.4888	–	–	–	–	–	0.4926	–
SJE [24]	2015	0.7458	0.3312	0.3395	0.6388	0.3381	–	–	–	–	–	0.4786	–
ESZSL [25]	2015	0.5722	0.3333	0.3951	0.3956	0.3333	–	–	–	–	–	0.4060	–
Feng(LSVM) [6]	2021	0.5868	0.5812	0.4451	0.5334	0.4256	–	–	–	–	–	0.5144	–
Feng(RF) [6]	2021	0.8840	0.5187	0.4833	0.6173	0.4312	–	–	–	–	–	0.5869	–
Feng(NB) [6]	2021	0.8027	0.6263	**0.6756**	0.7243	**0.6743**	–	–	–	–	–	0.6786	–
SCE(LSVM) [26]	2023	0.8970	0.5720	0.4760	0.6280	–	–	–	–	–	–	0.6420	–
SCE(RF) [26]	2023	**0.9090**	0.5730	0.5560	0.6710	–	–	–	–	–	–	0.6710	–
SCE(NB) [26]	2023	0.8950	**0.7810**	0.6240	**0.7600**	–	–	–	–	–	–	**0.7650**	–
Zhou et al. [1]	2021	–	–	–	–	–	–	–	–	–	–	0.6960	–
Kmeans [27]	2003	0.6736	0.5341	0.4228	0.4806	0.6277	0.4640	0.4278	0.3664	0.4792	0.4804	0.5091	0.4957
The proposed	2023	**0.9264**	**0.8549**	**0.6792**	**0.7388**	**0.6832**	**0.5674**	**0.6889**	**0.7444**	**0.6729**	**0.7229**	**0.7765**	**0.7279**

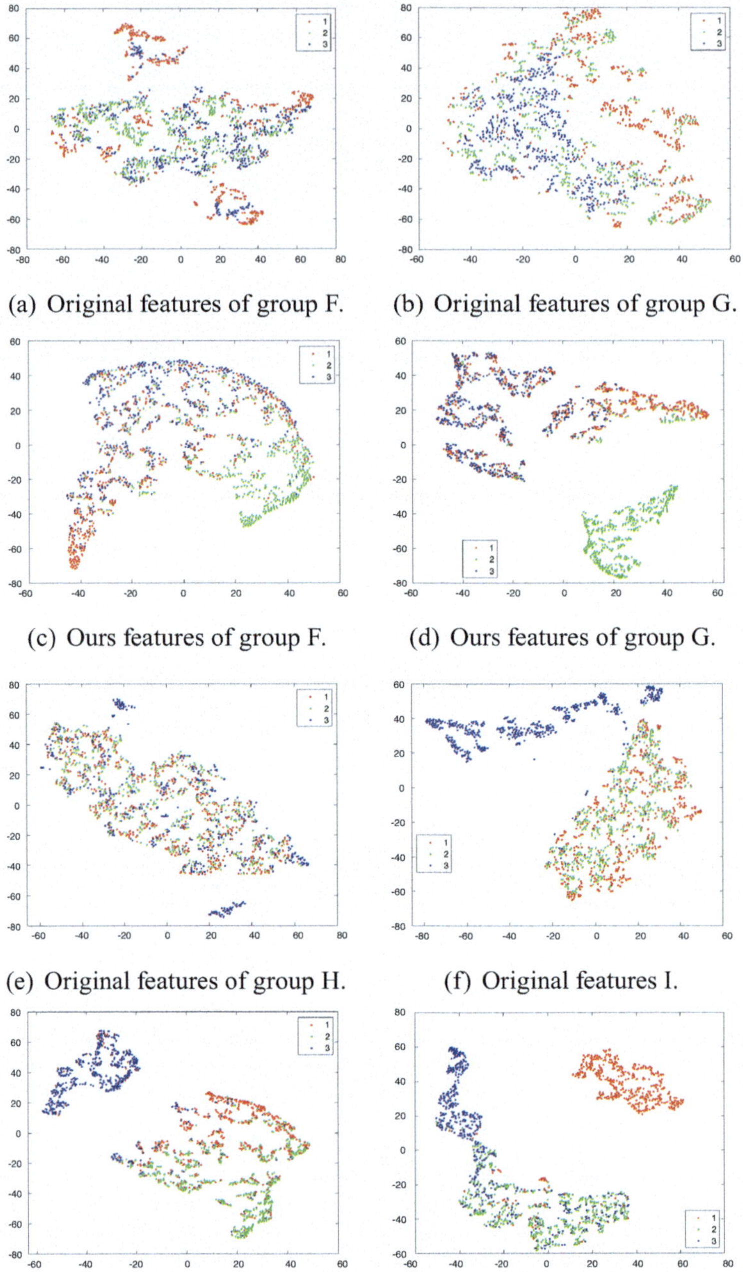

(a) Original features of group F. (b) Original features of group G.

(c) Ours features of group F. (d) Ours features of group G.

(e) Original features of group H. (f) Original features I.

(g) Ours features of group H. (h) Ours features of group I.

Fig. 8.3 The visualization of target fault features across eight groups is shown. The features in **a**, **b**, **e**, **f** are derived from the original space, while the features in **c**, **d**, **g**, and **h** are extracted using our methods

8.4.1.3 Convergence Analysis

In this section, the convergence of seven variables is examined with an increasing number of iterations. Figure 8.4a–g empirically demonstrates the speed at which the proposed algorithm attains convergence, where the x-axis represents the iteration number and the y-axis indicates the error value for each variable. The coefficients for the source domain G_S (Fig. 8.4a), the coefficients for the target domain G_T (Fig. 8.4e), and the knowledge dictionary D (Fig. 8.4d) converge after six iterations. The multi-class projection space for the source domain M_S (Fig. 8.4b) and the target domain M_T (Fig. 8.4f) converge at approximately the fifteenth iteration, while the learned features of X_S and X_T converge by the twentieth iteration. Additionally, we plot the accuracy over 50 iterations (Fig. 8.4h), showing that the accuracy stabilizes after 20 iterations.

8.4.2 Hot Steel Roll Process

8.4.2.1 Data Description

Then, we use data collected from hot steel roll process to evaluate the effectiveness of the proposed method. In the experiment, three categories are randomly selected from the dataset to serve as the source domain data, while the residual categories are designated as the target domain data. This process of random selection and domain separation is repeated 10 times to ascertain the credibility and robustness of the experimental results.

8.4.2.2 Results and Analysis

The fault identification results of the SCR dataset were compared against the k-means method as part of the ablation experiment. The introduced approach achieved a fault identification accuracy of 0.7150, while the k-means method only reached an accuracy of 0.5475. Our methodology is demonstrated to surpass the baseline by a notable margin of 0.1675, indicating a substantial improvement.

Additionally, we conducted a visualization experiment using t-SNE on a randomly selected set of target domain data. As shown in Fig. 8.5, (a) depicts the visualization of the original features, while (b) illustrates the feature visualization extracted by our proposed method. Despite having only 20 samples per class, it is clear that in Fig. 8.5a, the two classes are somewhat intermixed, whereas in Fig. 8.5b, the two classes are distinctly separated into two clusters. This confirms the effectiveness of our approach in terms of utilizing source domain knowledge to derive distinctive and resilient features from zero-sample data in the target domain, ultimately enhancing the precision of fault detection and identification in the target domain.

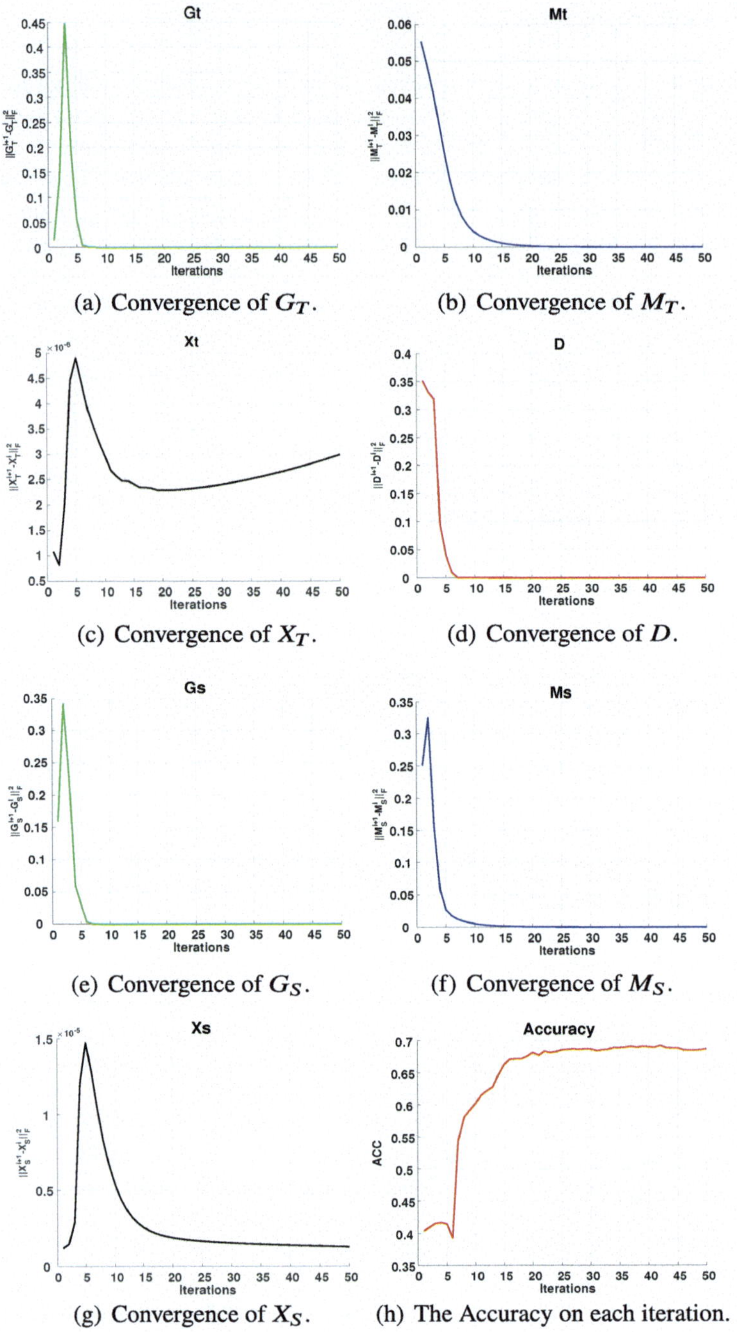

(a) Convergence of G_T. (b) Convergence of M_T.

(c) Convergence of X_T. (d) Convergence of D.

(e) Convergence of G_S. (f) Convergence of M_S.

(g) Convergence of X_S. (h) The Accuracy on each iteration.

Fig. 8.4 Convergence of seven variables on 50 iterations as well as the accuracy

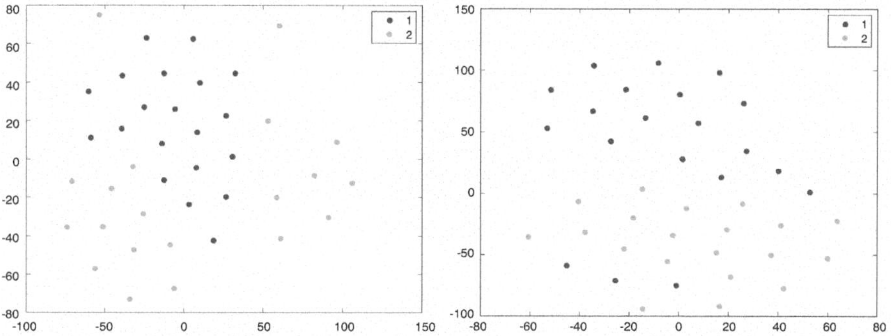

(a) Original features. (b) Our features.

Fig. 8.5 The visualization of features of target faults on the SCR dataset, where **a** is features that come from original space, and **b** is the feature extracted by our methods

8.5 Conclusion

This chapter presents a novel ZSFD method based on transfer learning. Firstly, a multi-class space projection model was introduced to address the adverse impact between different classes. Next, a shared fundamental knowledge dictionary was extracted from source data and transferred to target data to enhance fault identification accuracy. Additionally, a pseudo-label mechanism was implemented to explore the inter-class and intra-class information of target data. this chapter provides an innovative perspective on ZSFD through feature learning and shared knowledge transfer. Experiments on the TEP and a real hot roll of steel process demonstrate that the proposed methodology surpasses existing techniques in performance, even though the baseline in this chapter is a semi-supervised method. Furthermore, the proposed multi-space strategy shows versatility, suggesting its potential application for feature extraction in various other methodologies. As a potential future research direction, further exploration of feature extraction using the multi-space strategy is likely to yield significant insights and contribute to advancing this approach.

References

1. Zhuo Y, Ge Z (2021) Auxiliary information-guided industrial data augmentation for any-shot fault learning and diagnosis. IEEE Trans Indus Inform 17(11):7535–7545
2. Sun C, Ma M, Zhao Z, Tian S, Yan R, Chen X (2018) Deep transfer learning based on sparse autoencoder for remaining useful life prediction of tool in manufacturing. IEEE Trans Indus Inform 15(4):2416–2425
3. Lampert CH, Nickisch H, Harmeling S (2009) Learning to detect unseen object classes by between-class attribute transfer. In: 2009 IEEE conference on computer vision and pattern recognition, pp 951–958. IEEE

4. Lampert CH, Nickisch H, Harmeling S (2013) Attribute-based classification for zero-shot visual object categorization. IEEE Trans Pattern Anal Mach Intell 36(3):453–465
5. Mensink T, Gavves E, Snoek CGM (2014) Costa: Co-occurrence statistics for zero-shot classification. In: Proceedings of the IEEE conference on computer vision and pattern recognition, pp 2441–2448
6. Feng L, Zhao C (2020) Fault description based attribute transfer for zero-sample industrial fault diagnosis. IEEE Trans Indus Inform 17(3):1852–1862
7. Xu J, Long Z, Weihua Z, Fan Y, Ding X, Yuan X (2022) Zero-shot learning for compound fault diagnosis of bearings. Expert Syst Appl 190:116197
8. Xing S, Lei Y, Wang S, Na L, Li N (2022) A label description space embedded model for zero-shot intelligent diagnosis of mechanical compound faults. Mech Syst Signal Process 162:108036
9. Wang Z, He X, Yang B, Li N (2021) Subdomain adaptation transfer learning network for fault diagnosis of roller bearings. IEEE Trans Indus Electron 69(8):8430–8439
10. Chai Z, Zhao C, Huang B (2021) Multisource-refined transfer network for industrial fault diagnosis under domain and category inconsistencies. IEEE Trans Cybern 52(9):9784–9796
11. Xu J, Long Z, Weihua Z, Fan Y, Ding X, Yuan X (2022) Zero-shot learning for compound fault diagnosis of bearings. Expert Syst Appl 190:116197
12. Chen Z, Jun W, Deng C, Wang X, Wang Y (2022) Deep attention relation network: a zero-shot learning method for bearing fault diagnosis under unknown domains. IEEE Trans Reliab 72(1):79–89
13. Jiao J, Zhao M, Lin J, Liang K (2020) Residual joint adaptation adversarial network for intelligent transfer fault diagnosis. Mech Syst Signal Process 145:106962
14. Sun M, Wang H, Liu P, Huang S, Wang P, Meng J (2021) Stack autoencoder transfer learning algorithm for bearing fault diagnosis based on class separation and domain fusion. IEEE Trans Indus Electron 69(3):3047–3058
15. Qin Y, Qian Q, Luo J, Huayan P (2022) Deep joint distribution alignment: a novel enhanced-domain adaptation mechanism for fault transfer diagnosis. IEEE Trans Cybern 53(5):3128–3138
16. Li C, Li S, Wang H, Gu F, Ball AD (2023) Attention-based deep meta-transfer learning for few-shot fine-grained fault diagnosis. Knowl Based Syst 264:110345
17. Wang Y, Yan J, Ye X, Jing Q, Wang J, Geng Y (2022) Few-shot transfer learning with attention mechanism for high-voltage circuit breaker fault diagnosis. IEEE Trans Indus Appl 58(3):3353–3360
18. Li S, Shao M, Yun F (2017) Person re-identification by cross-view multi-level dictionary learning. IEEE Trans Pattern Anal Mach Intell 40(12):2963–2977
19. Pan SJ, Yang Q (2009) A survey on transfer learning. IEEE Trans Knowl Data Eng 22(10):1345–1359
20. Lee D-H et al. (2013) Pseudo-label: the simple and efficient semi-supervised learning method for deep neural networks. In: Workshop on challenges in representation learning, ICML, vol 3, page 896. Atlanta
21. Castelli V, Cover TM (1996) The relative value of labeled and unlabeled samples in pattern recognition with an unknown mixing parameter. IEEE Trans Inform Theory 42(6):2102–2117
22. Lampert CH, Nickisch H, Harmeling S (2009) Learning to detect unseen object classes by between-class attribute transfer. In: 2009 IEEE conference on computer vision and pattern recognition, pp 951–958. IEEE
23. Lampert CH, Nickisch H, Harmeling S (2013) Attribute-based classification for zero-shot visual object categorization. IEEE Trans Pattern Anal Mach Intell 36(3):453–465
24. Akata Z, Reed S, Walter D, Lee H, Schiele B (2015) Evaluation of output embeddings for fine-grained image classification. In: Proceedings of the IEEE conference on computer vision and pattern recognition, pp 2927–2936
25. Romera-Paredes B, Torr P (2015) An embarrassingly simple approach to zero-shot learning. In: International conference on machine learning, pp 2152–2161. PMLR

26. Zhengwei H, Zhao H, Yao L, Peng J (2022) Semantic-consistent embedding for zero-shot fault diagnosis. IEEE Trans Indus Inform 19(5):7022–7031
27. Likas A, Vlassis N, Verbeek JJ (2003) The global k-means clustering algorithm. Pattern Recogn 36(2):451–461

Chapter 9
Incremental Conditional Drift Adaptation for Processes Fault Identification Under Non-stationary Working Conditions

For non-stationary industrial processes, the distribution of data often deviates from historical patterns due to various factors, such as the aging of equipment and adjustments in manufacturing strategies. As a result, the conventional methods, which presume an identical and independently distributed (i.i.d.) setting, may falter in delivering satisfactory diagnostic results. This chapter focuses on the phenomenon of conditional drift, a relaxation from the stringent i.i.d. assumption, indicating that the conditional distribution of industrial data evolves over time. From a mathematical lens, we initially formulate the issue of conditional drift and propose a target mapping approach for drift adaptation, grounded in the principle of minimizing risk. Adopting this strategy, we devise an adaptive incremental identification model. This methodology enables our approach to adapt to conditional drift, enhancing its resilience in dynamic settings. The efficacy of our proposed method is substantiated through fault identification applications in two industrial processes.

9.1 Motivation

Batch learning based fault identification techniques fundamentally involve utilizing all labeled data from established fault classes to construct a classifier that subsequently predicts unlabeled industrial data. However, they are generally tailored for initial training datasets, whereas contemporary industrial processes generate substantial volumes of data on a continuous basis. Upon acquiring newly collected fault data, these methods necessitate re-training to update the model, resulting in substantial computational costs that can often be prohibitive.

Incremental learning presents a feasible solution to the above constraint by allowing prediction models to be progressively trained as labeled data is sequentially acquired, enabling predictions on newly incoming, unseen data [1]. While incremental models exhibit robust performance, they often assume that data collected at various times are independently and identically sampled from the same distribution.

H. Yin et al., *Data-Driven Fault Diagnosis for Complex Industrial Processes*,
Engineering Applications of Computational Methods 22,
https://doi.org/10.1007/978-981-96-3153-7_9

However, in non-stationary industrial settings, the distribution of new data frequently diverges from that of historical data, owing to factors such as equipment wear and environmental alterations. Consequently, the diagnostic models' decision boundaries undergo gradual shifts. While many approaches can adapt to changing environments through incremental learning mechanisms, their effectiveness and efficiency in recovering from distribution drift are not always guaranteed. Figure 9.1 depicts the diagnostic capabilities of various learning methods when subjected to the detrimental effects of drifting. In the presence of distribution drift, batch learning models experience a decline in performance and are unable to recover from degradation, as they lack the ability to integrate drift information from historical data. Conversely, incremental models possess the capability to continuously learn from newly acquired data, enabling them to gradually recover from performance dips and ultimately stabilize. We hypothesize that by incorporating drift adaptation strategies, it is possible to achieve faster distribution matching over time, enabling models to be more robust against changes, with shorter recovery times, smaller maximum performance drops, and improved steady-state performance. Hence, it is crucial to develop a fault identification model that possesses the capability to adapt to drift in evolving industrial settings.

$$P(\mathbf{Y}|\mathbf{X}) = \frac{P(\mathbf{Y})P(\mathbf{X}|\mathbf{Y})}{\sum_{i=1}^{N} P(y_i)P(\mathbf{X}|y_i)} \tag{9.1}$$

By Bayes' rule, the diagnostic decision, represented as the posterior probability $P(\mathbf{Y}|\mathbf{X})$, where $\mathbf{X}$ stands for industrial data and $\mathbf{Y}$ for fault labels, is determined by both the prior probability $P(\mathbf{Y})$ and the class-conditional probability $P(\mathbf{X})$. Recently, research has increasingly examined target drift in fault identification, a phenomenon

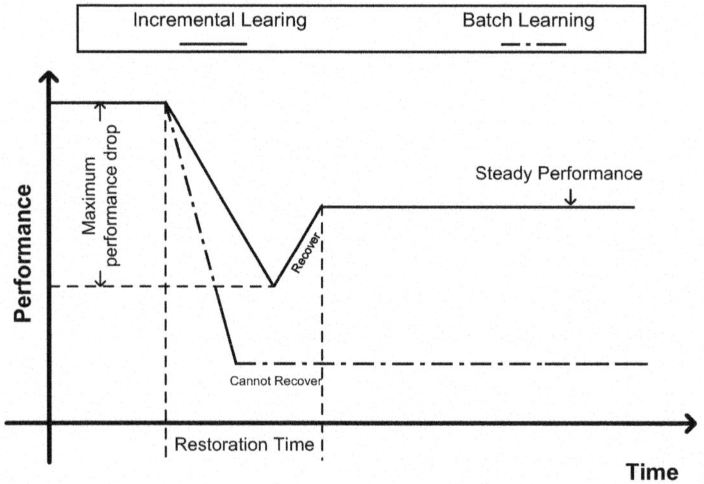

Fig. 9.1 The identification performance with different learning strategy

where the marginal distribution of the target (fault classes) varies over time. Such drift is frequently observed in practice, for instance, with the occurrence of new faults [2] or with dynamic imbalances in fault classes [3]. Conversely, the latter drift, termed conditional drift, represents a practical challenge in which the class-conditional distribution of data changes over time. For example, in the case of an identical industrial fault, features such as means and variances may vary due to shifts in manufacturing strategies, differences in feedstock specifications, and other factors. Managing this type of variation allows for matching conditional distributions, which enhances the robustness of diagnostic models. This chapter primarily addresses fault identification under conditional drift, a more complex multi-class classification problem.

To tackle this issue, the chapter begins by analyzing conditional drift from a mathematical standpoint and presents a target mapping strategy to inform the design of diagnostic models. Building on this approach, an incremental identification method with adaptation to conditional drift is proposed. In particular, the diagnostic model includes both an offline classifier and an online classifier. The offline classifier is trained on historical data to retain previous knowledge and mitigate catastrophic forgetting, whereas the online classifier is continually updated with incoming data, adapting to changes in industrial data streams.The two classifiers are dynamically integrated using the Hedge strategy. Furthermore, a transformation matrix is constructed to minimize the distribution gap between historical and incoming data. This approach allows the proposed method to swiftly adapt to conditional drift, enhancing its robustness in a fluctuating environment.

The contributions of our approach are outlined below:

- This chapter provides a mathematical definition of conditional drift in fault identification and introduces a target-mapping-based minimum risk criterion to guide the design of diagnostic models.
- This chapter presents a diagnostic model capable of incremental learning and adaptation to conditional drift. Specifically, a transformation matrix is crafted to continuously address the distribution discrepancies resulting from drift.
- The proposed method is applied to two benchmark tests to demonstrate its effectiveness in adapting to conditional drift.

9.2 Problem Formulation

Conditional drift describes the phenomenon in which the statistical properties of streaming data shift unpredictably over time, resulting in a mismatch between the class-conditional distribution of historical and new data. In real-world industrial systems, this drift can stem from factors like equipment aging or environmental variations. Mathematically, conditional drift is defined as follows [4].

Over a time period $[0, t]$, consider a set of d-dimensional industrial data collected as $\mathcal{D} = (\mathbf{x}_1, y_1), (\mathbf{x}_2, y_2), \ldots, (\mathbf{x}_t, y_t)$, where each $\mathbf{x}_i$ is a data vector and y_i represents the corresponding label. The dataset $\mathbf{X}^t$ adheres to a distribution $P_t(\mathbf{Y}^t)$ and

$P_t(\mathbf{X}^t|\mathbf{Y}^t)$. Conditional drift may arise if the industrial environment changes. For a new dataset $\mathcal{D}^\Delta$ received at time $t + \Delta$, we have:

$$P_t(\mathbf{Y}^t) = P_{t+\Delta}(\mathbf{Y}^{t+\Delta}) \tag{9.2}$$
$$P_t(\mathbf{X}^t|\mathbf{Y}^t) \neq P_{t+\Delta}(\mathbf{X}^{t+\Delta}|\mathbf{Y}^{t+\Delta})$$

As a result of Bayes' rule, it is highly unlikely that $P_{t+\Delta}(\mathbf{Y}^{t+\Delta}|\mathbf{X}^{t+\Delta})$ remains unchanged over time:

$$P_t(\mathbf{Y}^t|\mathbf{X}^t) \neq P_{t+\Delta}(\mathbf{Y}^{t+\Delta}|\mathbf{X}^{t+\Delta}) \tag{9.3}$$

Figure 9.2 provides a clear illustration of conditional drift in a two-dimensional space. At time t, a decision boundary is established based on the current data to classify system conditions. As time advances, data features change, prompting a shift in the optimal decision boundary. If the trained model remains unchanged and is directly applied at time $t + \Delta$, the resulting decisions may yield suboptimal fault identification performance.provides a clear illustration of conditional drift in a two-dimensional space. At time t, a decision boundary is established based on the current data to classify system conditions. As time advances, data features change, prompting a shift in the optimal decision boundary. If the trained model remains unchanged and is directly applied at time $t + \Delta$, the resulting decisions may yield suboptimal fault identification performance.

To handle conditional drift, adaptive and continuous learning of mappings can be used to minimize distribution mismatches throughout the incremental learning process. Such mappings may be implemented through various techniques, including parameter mapping, source mapping, target mapping, and co-mapping [5]. This chapter presents a target mapping strategy, defined as follows:

$$P_t(\mathbf{Y}^t|\mathbf{X}^t; \theta^t_{\mathcal{F}}) = P_\Delta(\mathbf{Y}^\Delta|\mathcal{A}\mathbf{X}^\Delta; \theta^{t+\Delta}_{\mathcal{F}}) \tag{9.4}$$

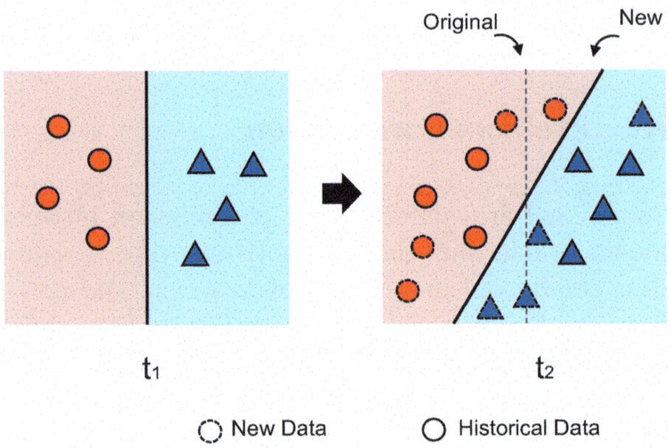

Fig. 9.2 The decision boundary impacted by the conditional drift phenomenon

Here, $\mathcal{A}$ represents a mapping, and $\theta^t_{\mathcal{F}}$ indicates the parameters of a fault identification model $\mathcal{F}$ at time t. The symbol $\mathbf{X}^{\Delta}$ refers to the batch of newly received training industrial data within the time interval $[t, t + \Delta]$. The target mapping aligns the distribution of new data to match that of historical data. One key advantage of this approach is that adaptation can occur after acquiring a previously well-trained fault identification model.

Typically, training a fault identification model over the interval $[0, t]$ aims to minimize the expected classification risk:

$$\min_{\theta^t_{\mathcal{F}} \in \Theta_{\mathcal{F}}} \mathbb{E}_{(\mathbf{x},y)\sim P_t}[\mathcal{L}(\mathbf{x}, y, \theta^t_{\mathcal{F}})] \tag{9.5}$$

where $\mathcal{L}(\mathbf{x}, y, \theta_{\mathcal{F}})$ is the loss function of a classifier depending on parameters $\theta^t_{\mathcal{F}}$. It could be a misclassification loss or a negative log-likelihood. Since $P_t(\mathbf{X}^t)$ is unknown, The empirical distribution $\widetilde{P}_t(\mathbf{X}^t)$ serves as an estimator for $P_t(\mathbf{X}^t)$. Consequently, Eq. (9.5) can be reformulated as:

$$\min_{\theta^t_{\mathcal{F}} \in \Theta_{\mathcal{F}}} \frac{1}{t} \sum_{i=1}^{t} [\mathcal{L}(\mathbf{x}_i, y_i, \theta^t_{\mathcal{F}})] \tag{9.6}$$

In $[t, t + \Delta]$, we wish to generalize the above fault identification model by Eq. (9.4), in order to suit new data draw from $P_{\Delta}(\mathbf{X}^{\Delta}, y^{\Delta})$: Over the interval $[t, t + \Delta]$, we aim to extend the above fault identification model using Eq. (9.4) to accommodate new data drawn from $P_{\Delta}(\mathbf{X}^{\Delta}, y^{\Delta})$:

$$\begin{aligned} \mathbb{E}_{(\mathbf{x},y)\sim P_t}[\mathcal{L}(\mathbf{x}, y, \theta^t_{\mathcal{F}})] &= \mathbb{E}_{(\mathbf{x},y)\sim P_{\Delta}}[\mathcal{L}(\mathcal{A}\mathbf{x}, y, \theta^{t+\Delta}_{\mathcal{F}})] \\ &= \frac{1}{N^{\Delta}} \sum_{i=1}^{N^{\Delta}} [\mathcal{L}(\mathcal{A}\mathbf{x}_i, y_i, \theta^{t+\Delta}_{\mathcal{F}})] \end{aligned} \tag{9.7}$$

where N^{Δ} is the number of data in the time period $[t, t + \Delta]$.

As indicated in Eq. (9.7), the risk expectation for an updated classifier on $(\mathbf{x}, y) \sim P_{\Delta}$ at time $t + \Delta$ is equivalent to that of a historical classifier on $(\mathbf{x}, y) \sim P_t$ at time t. Continuously optimizing this objective addresses the online distribution adaptation problem. The primary challenge involves designing an effective classifier and adaptively selecting the transformation matrix.

9.3 Methodology

This section presents the proposed incremental fault identification approach, equipped with the capability to adapt to conditional drift. We begin by outlining the method's overall framework.

9.3.1 The Overall Framework

Drawing inspiration from [6, 7], the overall fault identification framework is structured as shown in Fig. 9.3. This framework includes two main stages: offline training and online updating.

Offline Training Stage ($i \in (0, t]$): Initially, an initial transformation matrix $\mathcal{A}$ is derived. Historical data $\mathbf{x}_i \in \mathbf{X}^h$ are subsequently projected into a new feature space via $\mathcal{A}\mathbf{x}_i$ and used to train a historical classifier $\mathcal{F}^h$.

Online Updating Stage ($i \in (t, +\infty]$): When a new training sample $\mathbf{x}_i$ becomes available, it is transformed via $\mathcal{A}\mathbf{x}_i$ and used to update an online classifier $\mathcal{F}^o$. The diagnostic model $\mathcal{F}$ is then constructed by combining the historical classifier $\mathcal{F}^h$ with the online classifier $\mathcal{F}^o$. The transformation matrix $\mathcal{A}$ is also updated accordingly.

In fault identification, an unlabeled industrial data point is input into the ensemble classifier $\mathcal{F}$ to assess whether the system is operating normally or has a fault.

9.3.2 Offline Training Stage

Transformation Matrix Initialization

During the offline stage, as the historical dataset collected over different time periods may be influenced by conditional drift, we project the data into a new feature space to minimize distributional differences. This process aids in determining the initial transformation matrix $\mathcal{A}$.

Specifically, the historical dataset $\mathbf{X}^h$ is initially divided into two subsets based on time, such that $\mathbf{X}^h = [\mathbf{X}^{h_1}, \mathbf{X}^{h_2}]$. Using Joint Domain Adaptation (JDA) [8], we

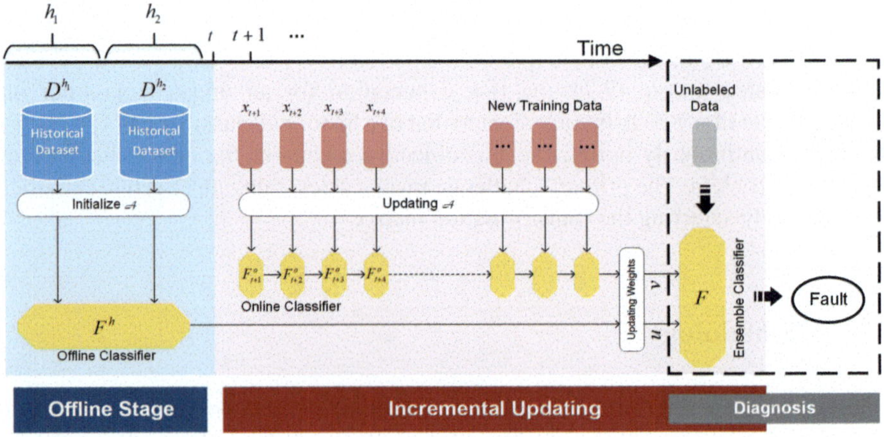

Fig. 9.3 The overall framework of the proposed method

map these two subsets, which originate from different distributions, into a feature space where both the marginal and conditional distributions are minimized.

$$\min_{\mathcal{A}} ||P_{h_1}(\mathcal{A}\mathbf{X}^{h_1}) - P_{h_2}(\mathcal{A}\mathbf{X}^{h_2})|| + ||P_{h_1}(\mathbf{Y}^{h_1}|\mathcal{A}\mathbf{X}^{h_1}) - P_{h_2}(\mathbf{Y}^{h_2}|\mathcal{A}\mathbf{X}^{h_2})|| \tag{9.8}$$

Estimating $P(y|\mathbf{x})$ is, unfortunately, quite complex. Therefore, we approximate $P(y|\mathbf{x})$ using the class-conditional distribution $P(\mathbf{x}|y)$. Consequently, Eq. (9.8) is computed as follows:

$$\left\| \frac{1}{h_1} \sum_{j=1, \mathbf{x}_j \in \mathcal{D}^{h_1}}^{h_1} \mathcal{A}\mathbf{x}_j - \frac{1}{h_2} \sum_{j=1, \mathbf{x}_j \in \mathcal{D}^{h_2}}^{h_2} \mathcal{A}\mathbf{x}_j \right\|_{\mathcal{H}} + \sum_{c=1}^{C} \left\| \frac{1}{h_1^{(c)}} \sum_{\mathbf{x}_j \in \mathcal{D}_{(c)}^{h_1}} \mathcal{A}\mathbf{x}_j - \frac{1}{h_2^{(c)}} \sum_{\mathbf{x}_j \in \mathcal{D}_{(c)}^{h_2}} \mathcal{A}\mathbf{x}_j \right\|_{\mathcal{H}} \tag{9.9}$$

where $\mathcal{H}$ denotes reproducing kernel Hilbert space. h_1 is the data number $\mathbf{X}^{h_1}$, i.e. $h_1 = |\mathbf{X}^{h_1}|$. C is the number of data classes. $\mathbf{X}_c^{h_1} = \{(\mathbf{x}_i, y_i) \in \mathbf{X}^{h_1} \wedge y_i = c\}$. $h_1^{(c)} = |\mathbf{X}_{(c)}^{h_1}|$.

Solving Eq. (9.9) provides the initial transformation matrix $\mathcal{A}$. However, this problem is challenging. Kernel methods, as discussed in [8], can be employed to compute $\mathcal{A}$. For further details on the optimization process, readers are encouraged to consult that work.

After obtaining $\mathcal{A}$, each $\mathbf{x}_i \in \mathbf{X}^t$ is projected into a new feature space as $\mathcal{A}\mathbf{x}_i$. This mapping provides two primary benefits: (1) it reduces the dimensionality of high-dimensional data, thereby lowering the computational cost of the subsequent learning process; and (2) it aligns the distribution of historical data across different time periods.

Historical Classifier Learning

In this chapter, we employ the Passive Aggressive (PA) algorithm [9] as the component classifier. The proposed algorithm functions as a general framework, allowing for other incremental learning algorithms to be used as base learners as well.

Specifically, the historical classifier $\mathcal{F}^h$ is trained on a sequence of historical data $(\mathcal{A}\mathbf{x}_i, y_i)$, where $i \in (0, t]$. The PA algorithm then constructs a set of weight vectors:

$$\mathcal{F}^{h_i} = [\mathbf{f}_i^{h,(1)}, \ldots, \mathbf{f}_i^{h,(c)}, \ldots, \mathbf{f}_i^{h,(C)}] \quad where \ \mathbf{f}_i^{h,(c)} = \mathbf{w}_i^{(c)} \cdot \mathcal{A}\mathbf{x}_i \tag{9.10}$$

At time i, upon receiving a training sample $\mathcal{A}\mathbf{x}_i$, we compute the scores for all classes and make a prediction as follows:

Algorithm 1 Offline Training based on PA.

Input: Historical Data $\mathbf{X}^h$ and labels, Number of classes C, Classifier parameter R.
Output: Historical Classifier $\mathcal{F}^h$.
1: Dividing $\mathbf{X}^h$ into two subsets
2: Initialize the transformation matrix $\mathcal{A}$ by Eq. (9.9).
3: **for** $i \leq t$ **do**
4: Mapping $\mathcal{A}\mathbf{x}_i^h$.
5: Training the classifier $\mathcal{F}_i^h$ by Eq. (9.14).
6: **end for**

$$\hat{y}_t = \arg\max_r \mathbf{f}_i^{h,(r)} = \mathbf{w}_i^{(r)} \cdot \mathcal{A}\mathbf{x}_i \tag{9.11}$$

Next, the true label y_t is used to compute the loss value using the following function:

$$\mathcal{L}_i(\mathcal{F}^{h_i}, \mathbf{x}_i, y_i) = \begin{cases} 0, & \mathbf{w}_i^{(y_i)}\mathbf{x}_i - \mathbf{w}_i^{(s_i)} \cdot \mathbf{x}_i \geq 1 \\ 1 - \mathbf{w}_i^{(y_i)}\mathbf{x}_i + \mathbf{w}_i^{(s_i)}\mathbf{x}_i, & otherwise \end{cases} \tag{9.12}$$

where $s_i = \arg\max_{s_i \neq y_i} \mathbf{f}_i^{h,(s_i)}$ represents the irrelevant class for $\mathbf{x}_i$ with the highest predicted score. The hinge loss ensures both a positive margin and dependable relative ranking.

Consequently, the updated vector $\mathbf{w}_{i+1}^{c}$ is obtained by solving the following problem:

$$\mathbf{w}_{i+1}^{(c)} = \arg\min_{\mathbf{w}} \frac{1}{2} \left\| \mathbf{w}^{(c)} - \mathbf{w}_i^{(c)} \right\|^2 \quad s.t.\ \mathcal{L}_i(\mathbf{f}_i^{h,(c)}, \mathbf{x}_i, y_i) = 0 \tag{9.13}$$

It learns a new weight vector with minimal adjustment while maintaining a reliable score difference between classes y_i and s_i for the i-th example.

This leads to the following update:

$$\mathbf{w}_{i+1}^{(y_i)} = \mathbf{w}_i^{(y_i)} + \tau_i \mathcal{A}\mathbf{x}_i, \quad \mathbf{w}_{i+1}^{(s_i)} = \mathbf{w}_i^{(s_i)} - \tau_i \mathcal{A}\mathbf{x}_i \tag{9.14}$$

where there are three choices for τ_i:

$$\tau_i = \begin{cases} \frac{\mathcal{L}_i(\mathcal{F}_i^h, \mathbf{x}_i, y_i)}{\|\mathbf{x}_i\|^2} & \\ \min\{R, \frac{\mathcal{L}_i(\mathcal{F}_i^h, \mathbf{x}_i, y_i)}{\|\mathbf{x}_i\|^2}\} & (I) \\ \frac{\mathcal{L}_i(\mathcal{F}_i^h, \mathbf{x}_i, y_i)}{\|\mathbf{x}_i\|^2 + \frac{1}{2R}} & (II) \end{cases} \tag{9.15}$$

Here, R is a positive parameter controlling aggressiveness. We refer to the strategies as Prototype, Type I, and Type II, respectively.

The offline training procedure is outlined in Algorithm 1. The trained classifier $\mathcal{F}^h$ learns and retains the knowledge contained within the historical data.

9.3.3 *Incremental Updating Stage*

Online Classifier Learning

At the start of the incremental updating stage ($i = t$), an online classifier $\mathcal{F}_i^o$ is initialized using $\mathcal{F}^h$. Upon receiving a training sample $(\mathbf{x}_i, y_i)$, where $i \in (t, +\infty)$, $\mathcal{F}_i^o$ is updated according to Eq. (9.16).

Ensemble Classifier Learning

After each update of the online classifier $\mathcal{F}^o$, it is combined with the historical classifier $\mathcal{F}^h$, resulting in the following ensemble classifier:

$$\mathcal{F}_i = \mu_i \mathcal{F}^h + \upsilon_i \mathcal{F}_i^o \tag{9.16}$$

Here, μ_i and υ_i are weighting parameters with $\mu_i + \upsilon_i = 1$. The Hedge(β) algorithm [10] is used to dynamically update the weights μ_i and υ_i:

$$\begin{aligned} \mu_{i+1} &= \frac{\mu_i \beta^{\xi_i^h}}{\mu_i \beta^{\xi_i^h} + \upsilon_i \beta^{\xi_i^o}} \\ \upsilon_{i+1} &= \frac{\upsilon_i \beta^{\xi_i^o}}{\mu_i \beta^{\xi_i^h} + \upsilon_i \beta^{\xi_i^o}} \end{aligned} \tag{9.17}$$

where β is a decay parameter within the range (0, 1). The terms ξ_i^h and ξ_i^o represent designated misclassification values for classifiers F_i^h and F_i^o, respectively, which determine the degree of decay. They are defined as follows:

$$\xi_i^h = \begin{cases} 0 \; if \; \hat{y}_i^h \neq y_i \\ 1 \; if \; \hat{y}_i^h = y_i \end{cases}, \quad \xi_i^o = \begin{cases} 0 \; if \; \hat{y}_i^o \neq y_i \\ 1 \; if \; \hat{y}_i^o = y_i \end{cases} \tag{9.18}$$

Here, Eq. (9.17) suggests that higher misclassification values reduce the importance of the corresponding classifier.

Transformation Matrix Updating

To accommodate conditional drift, the transformation matrix must also be updated in an online manner.

Following Eq. (9.9), we update $\mathcal{A}$ to align the distributions of the two datasets, $\mathbf{X}^i$ and $\mathbf{X}^\Delta$, by jointly minimizing both the marginal and conditional distributions:

$$\begin{aligned} &\min_{\mathcal{A}} ||P_i(\mathbf{X}^i) - P_\Delta(\mathcal{A}\mathbf{X}^\Delta)|| \\ &\quad + ||P_i(\mathbf{Y}^t|\mathbf{X}^i) - P_\Delta(\mathbf{Y}^\Delta|\mathcal{A}\mathbf{X}^\Delta)|| \end{aligned} \tag{9.19}$$

Here, $\mathbf{X}^i$ represents all historical industrial data recorded up to time instant i, while $\mathbf{X}^\Delta$ denotes the batch of newly received training data between time i and $i + \Delta$. Consequently, the optimization problem for updating $\mathcal{A}$ is formulated as:

$$\arg\{\min_{\mathcal{A}} ||\mathcal{A} - \mathcal{A}_i||^{2_F} + \gamma \sum_{c=0}^{C} ||\bar{\mathbf{m}}^i_{(c)} - \mathcal{A}\bar{\mathbf{m}}^{\Delta}_{(c)}||_2^2\} \tag{9.20}$$

where γ is a positive trade-off parameter. When $c = 0$, $\bar{\mathbf{m}}^i_0$ and $\bar{\mathbf{m}}^{\Delta}_0$ are the marginal distributions $P_i(\mathbf{X}^i)$ and $P_{\Delta}(\mathbf{X}^{\Delta})$, respectively:

$$\bar{\mathbf{m}}^i_{(0)} = \frac{1}{i} \sum_{j=1,\mathbf{x}_j \in \mathcal{D}^i}^{i} \mathcal{A}_i \cdot \mathbf{x}_j, \quad \bar{\mathbf{m}}^{\Delta}_{(0)} = \frac{1}{\Delta} \sum_{j=1,\mathbf{x}_j \in \mathcal{D}^{\Delta}}^{\Delta} \mathbf{x}_j \tag{9.21}$$

When $c > 0$, $\bar{\mathbf{m}}^i_{(c)}$ and $\bar{\mathbf{m}}^{\Delta}_{(c)}$ are the class conditional distributions $P_i(\mathbf{Y}^i|\mathbf{X}^i)$ and $P_{\Delta}(\mathbf{Y}^{\Delta}|\mathbf{X}^{\Delta})$, respectively:

$$\bar{\mathbf{m}}^i_{(c)} = \frac{1}{i^c} \sum_{\mathbf{x}_j \in \mathcal{D}^i_{(c)}} \mathcal{A}_i \cdot \mathbf{x}_j, \quad \bar{\mathbf{m}}^{\Delta}_{(c)} = \frac{1}{\Delta^c} \sum_{\mathbf{x}_j \in \mathcal{D}^{\Delta}_{(c)}} \mathbf{x}_j \tag{9.22}$$

By solving Eq. (9.20), we can get the updating rule for $\mathcal{A}$:

$$\mathcal{A}_{i+1} \leftarrow (\mathcal{A}_i + \gamma \sum_{c=0}^{C} \bar{\mathbf{m}}^i_{(c)} [\bar{\mathbf{m}}^{\Delta}_{(c)}]^T)(\mathbf{I} + \gamma \sum_{c=0}^{C} \bar{\mathbf{m}}^{\Delta}_{(c)} [\bar{\mathbf{m}}^{\Delta}_{(c)}]^T)^{\dagger} \tag{9.23}$$

Here, $\mathbf{I}$ denotes the identity matrix, and $\dagger$ represents the pseudo-inverse. Following this, the vectors $\bar{\mathbf{m}}^i$ and $\bar{\mathbf{m}}^i_{(c)}$ must also be updated to restore the new data $\mathbf{X}^{\Delta}$ as historical information:

$$\begin{aligned} \bar{\mathbf{m}}^i_{(0)} &\leftarrow \frac{i \cdot \bar{\mathbf{m}}^i_{(0)} + \Delta \cdot \bar{\mathbf{m}}^{\Delta}_{(0)}}{i + \Delta} \\ \bar{\mathbf{m}}^i_{(c)} &\leftarrow \frac{i^c \cdot \bar{\mathbf{m}}^i_{(c)} + \Delta^c \cdot \bar{\mathbf{m}}^{\Delta}_{(c)}}{i^c + \Delta^c} \end{aligned} \tag{9.24}$$

Updating $\mathcal{A}$ is computationally expensive due to the matrix inversion involved. To address this, a time window T_w is introduced to control the frequency of updates. Specifically, the transformation matrix $\mathcal{A}$ is updated once every time window. For instance, in the subsequent experiments, $\mathcal{A}$ is updated after receiving 20 labeled samples.

We summarize the procedure of incremental updating in Algorithm 2.

Algorithm 2 Incremental Updating based on PA.

Input: N^n new data $\mathbf{X}^n$ and labels, Weighting parameter μ_0, υ_0, Trade-off parameter γ, Control window size T_w;
Output: Ensemble classifier $\mathcal{F}$.
1: **for** $i \leq N^n$ **do**
2: Mapping $\mathcal{A}\mathbf{x}^n$.
3: Updating the online classifier $\mathcal{F}_i^o$ by Eq. (9.14).
4: Updating the ensemble weights μ_i, υ_i by Eq. (9.17).
5: Obtaining the ensemble classifier $\mathcal{F}_i$ by Eq. (9.16).
6: **if** Reaching T_w limit **then**
7: Updating $\mathcal{A}$ by Eq. (9.23).
8: **end if**
9: **end for**

9.3.4 Fault Identification

The core idea of our model relies on an ensemble learning framework, where $\mathcal{F}^h$ maintains the historical industrial process knowledge, and $\mathcal{F}^o$ is updated to adapt to environmental changes.

For each new sample $\mathbf{x}_*$, the ensemble classifier $\mathcal{F}_i$ generates the predicted label as follows:

$$\hat{y}_* = \arg\max_r \mathcal{F}_i^{(r)} \tag{9.25}$$

This predicted label determines whether the system is operating in a faulty state.

9.4 Case Study

This section assesses the effectiveness of the proposed fault identification model through two scenarios: the Tennessee Eastman process and an actual multi-phase flow facility.

9.4.1 Tennessee Eastman Process

Data Description

We first use TEP for evaluation. At any given time, the observation vector consists of 41 measured variables and 11 manipulated variables. In this chapter, 15 different faults are simulated to represent common system malfunctions, with each fault divided into 480 training samples and 800 testing samples. To simulate conditional drift, the first 240 training samples are treated as historical data, while the entire set of 800 testing samples is considered new data. The distributional discrepancy between the historical and new data is assessed using Maximum Mean Discrepancy (MMD)

Table 9.1 The 15 faults in the TEP and the distribution discrepancy measured by MMD

No.	Fault description	MMD Value
1	A/C feed ratio, B composition constant (stream 4)	1340.67
2	B composition, A/C ratio constant (stream 4)	2013.13
3	D feed temperature (stream 2)	28.61
4	Reactor cooling water inlet temperature	12.50
5	Condenser cooling water inlet temperature	266.90
6	A feed loss (stream 1)	33531.96
7	C header pressure loss—reduced availability (stream 4)	326.74
8	A, B, C feed composition (stream 4)	933.78
9	D feed temperature (stream 2)	72.23
10	C feed temperature (stream 4)	74.75
11	Reactor cooling water inlet temperature	21.62
12	Condenser cooling water inlet temperature	374.63
13	Reactor kinetics	50120.39
14	Reactor cooling water valve	195.09
15	Condenser cooling water valve	33.18

[11], and the results are presented in Table 9.1. Higher MMD values correspond to greater differences in distribution. The findings indicate that all faults exhibit some degree of distribution mismatch between the historical and new data, with faults 6 and 13 showing the most pronounced discrepancies. For validation purposes, the study focuses on 9 faults (1, 2, 5, 6, 7, 8, 12, 13, and 14) that have MMD values exceeding 100.

Next, we present a visual representation of the conditional drift phenomenon by projecting the high-dimensional fault data into a 3-dimensional subspace. Sequential plots for faults 7 and 12 are shown in Fig. 9.4. In these plots, the x-axis represents time, while the y and z-axes correspond to the 1st and 2nd principal components of the fault data, which are obtained via Principal Component Analysis (PCA). The 3D scatter plots are further mapped onto the x-y, x-z, and y-z planes, which are shown in green, blue, and purple, respectively. Historical data points are represented as black dots. For instance, in the case of fault 12, the distribution of the new data progressively shifts over time, becoming more distinct from the historical data distribution. A similar pattern of drift is also observable for other faults. This drift highlights the evolving nature of the fault data over time, indicating a mismatch between the historical and newly acquired data distributions.

Settings and Comparison

After numerically and visually analyzing the conditional drift phenomenon in the TEP, we proceeded to evaluate the fault identification performance by comparing our proposed method with other well-known algorithms. For this evaluation, we used a total of 2160 samples from 9 different faults as historical data, and 7200 samples

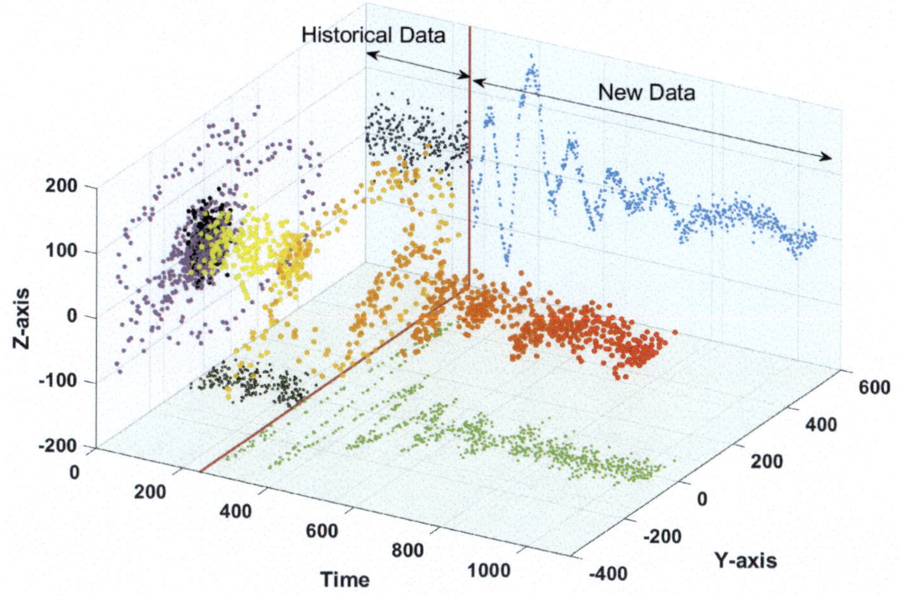

(a)

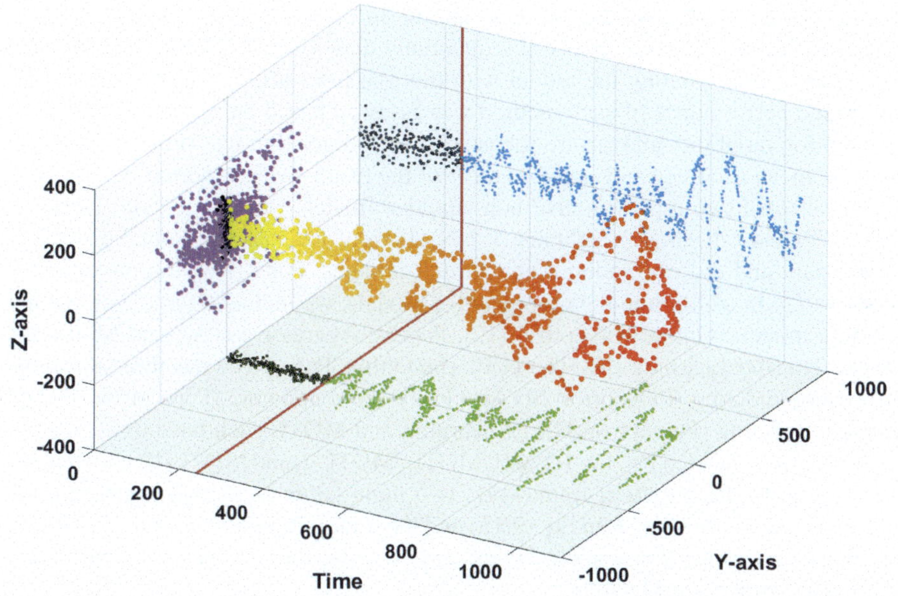

(b)

Fig. 9.4 The sequential plots of different faults in the 3-D sub-space. **a** The sequence of fault 7. **b** The sequence of fault 12

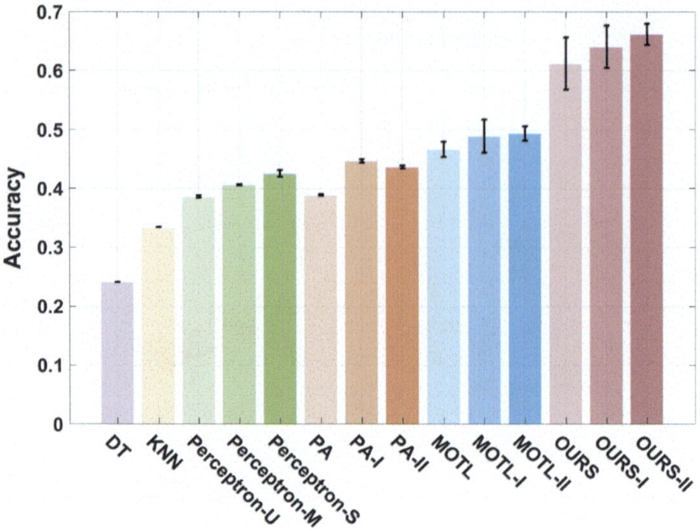

Fig. 9.5 The comparison results of different methods on the TEP

as new data. The evaluation process began with offline training using the historical data, followed by diagnosing the new data in an online manner. Specifically, the fault identification process involved sequentially diagnosing the fault label for each new sample and updating the model with the true label after each prediction. The diagnostic performance of each method was assessed based on accuracy (ACC). To account for variability and ensure robustness, each algorithm was run 20 times, and both the mean accuracy and standard error of the results were reported.

We selected two representative non-adaptive models for comparison: Decision Tree (DT) and k-Nearest Neighbors (k-NN), both of which were trained using historical data and did not undergo any further updates. For k-NN, the parameter k was set to 3. In addition to the non-adaptive models, we included three incremental learning models: Online Perceptron [12], Passive Aggressive (PA), and Multiclass Online Transfer Learning (MOTL) [13]. The Online Perceptron has three versions: max-score updating (Perceptron-M), similarity-score updating (Perceptron-S), and uniform updating (Perceptron-U). PA learning and MOTL each have three aggressiveness levels, labeled PA, PA-I, PA-II, MOTL, MOTL-I, and MOTL-II, with default parameters. In the proposed framework, two main parameters are considered: the aggressiveness parameter R in Eq. (9.15) and the trade-off parameter γ in Eq. (9.20). For this experiment, R was set to 10 and $\lg \gamma = 4$. Further details of the parameter study will be discussed later.

The comparison bar and the error curves are displayed in Figs. 9.5 and 9.6, respectively. Several noteworthy observations can be made from the results:

(1) Effectiveness of Incremental Models: Compared to non-adaptive models (KNN and DT), incremental models perform better in terms of accuracy. Non-adaptive

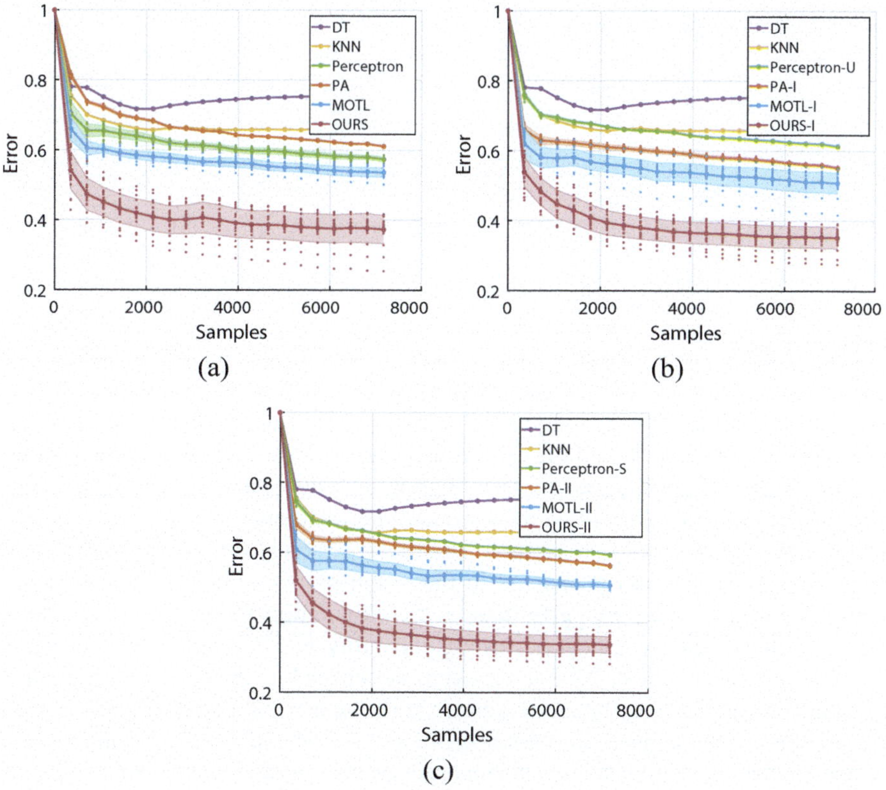

Fig. 9.6 The error curves of different identification models on the TEP. **a** Prototype, **b** Type I, **c** Type II

models are strictly limited to their applicability within initial training datasets and are unable to cope with the discrepancies in data distributions that arise in evolving environments. In contrast, incremental models can continuously learn from new samples, leading to improved diagnostic performance.

(2) Performance of the Proposed Method: Among the incremental models, the proposed algorithm achieves the highest accuracy, exceeding 60%. The distribution adaptation strategy is credited for this notable enhancement, facilitating swift matching of distributions across various time points and bolstering the model's resilience and consistent performance.

(3) Best Variant: Among the variants of the proposed method, OURS-II achieves the best performance with an accuracy of $67.90\% \pm 3.02\%$. In practical systems, the PA model, which adapts as required to fulfill optimization constraints, may be negatively impacted by noise or outliers. The aggressive parameter R in Type-II mitigates these negative impacts by relaxing the aggressive optimization, thereby improving performance.

Table 9.2 The ablation study settings

Strategy	#1	#2	#3	#4	#5	#6	#7	#8	#9
$\mathcal{F}^h$	√	√	√				√	√	√
$\mathcal{F}^o$	√	√	√	√	√	√			
Fixed $\mathcal{A}$		√			√			√	
Adaptive $\mathcal{A}$	√			√			√		

Ablation Study

On one hand, our framework dynamically integrates the historical classifier $\mathcal{F}^h$ with the online classifier $\mathcal{F}^o$, adapting to the evolving data. The historical classifier preserves past knowledge, while the online classifier adapts to newly incoming information. On the other hand, our framework progressively updates the transformation matrix $\mathcal{A}$ to mitigate distribution discrepancies between historical and new data. To demonstrate the effectiveness of these strategies, we conducted an ablation study, detailed in Table 9.2 and visualized in the violin plots of Fig. 9.7. For example, as shown in Fig. 9.7(c), the performance of configuration #6 is superior to that of #9, likely due to the incremental learning capability of $\mathcal{F}^o$, which better adapts to the evolving environment. Moreover, using the ensemble classifier (configuration #3) results in significantly better diagnostic performance compared to using a single classifier. From the comparisons between #2 and #3, #4 and #5, and #7 and #8, it is evident that incorporating the transformation matrix $\mathcal{A}$ improves performance significantly. Continuously updating $\mathcal{A}$ (configurations #1, #4, and #7) further enhances diagnostic accuracy. With the complete set of ensemble and adaptation strategies (configuration #1), our framework delivers the highest performance among all competitors, achieving an accuracy of 67.90% $\pm$ 3.02

Visualization Analysis

Next, we perform experiments to clearly demonstrate the effectiveness of our framework in addressing conditional drift. Specifically, we compare the data distributions with and without the application of our framework by visualizing them in both two-dimensional and one-dimensional spaces using an Auto-Encoder (AE) [14]. For example, we analyze fault 5. Figure 9.8 illustrates the original fault 5 data projected into the 2-D space, where darker dots indicate more recently acquired data. It is evident that the distribution of the historical data significantly diverges from that of the new data. In contrast, when applying our framework, the data distributions in the latent space become better aligned. As shown in Fig. 9.8 for the 1-D space, the datasets processed by our framework exhibit less separation. The probability density curves overlap more closely, and the mean difference between the historical and new data diminishes after adaptation.

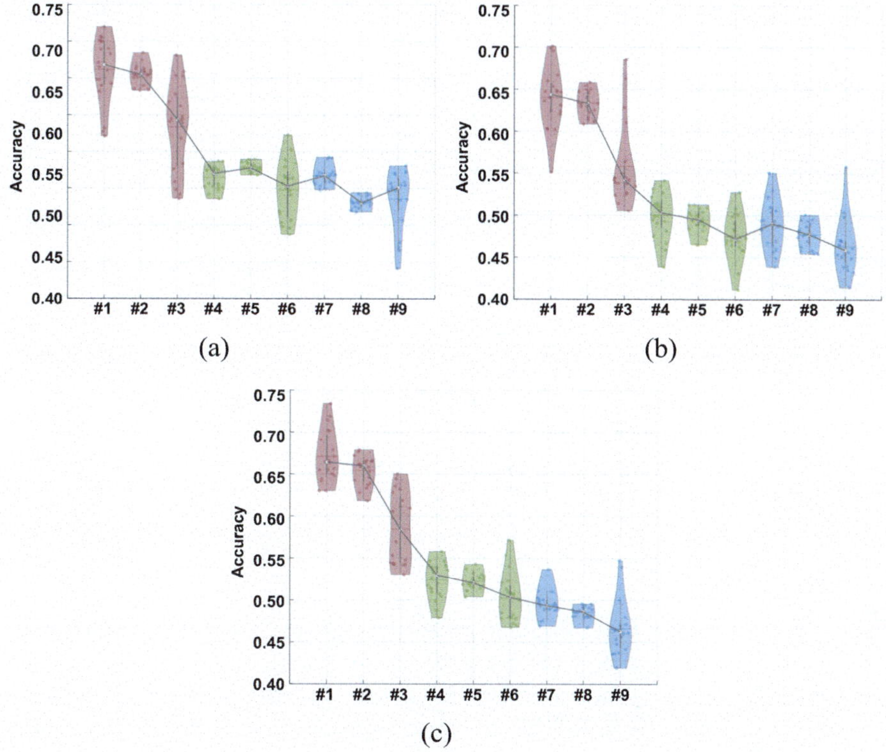

Fig. 9.7 Ablation study of the proposed method. **a** Prototype, **b** Type I, **c** Type II

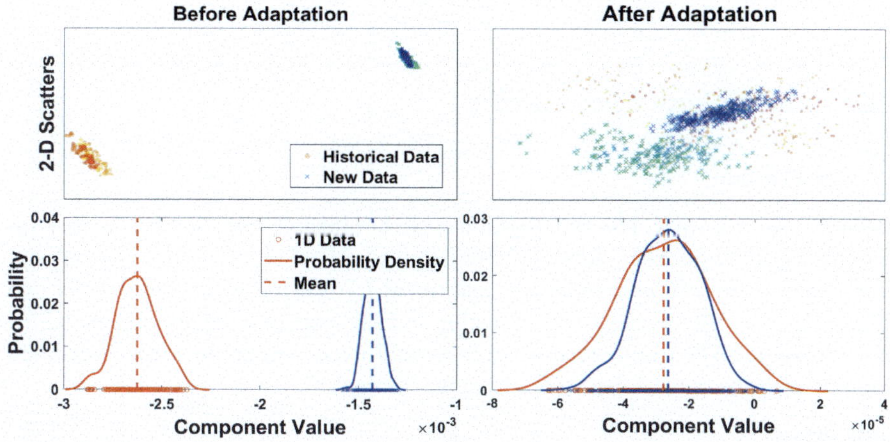

Fig. 9.8 The visualization of fault samples with/without adaptation in the1-D and 2-D sub-space by AE, on fault 5

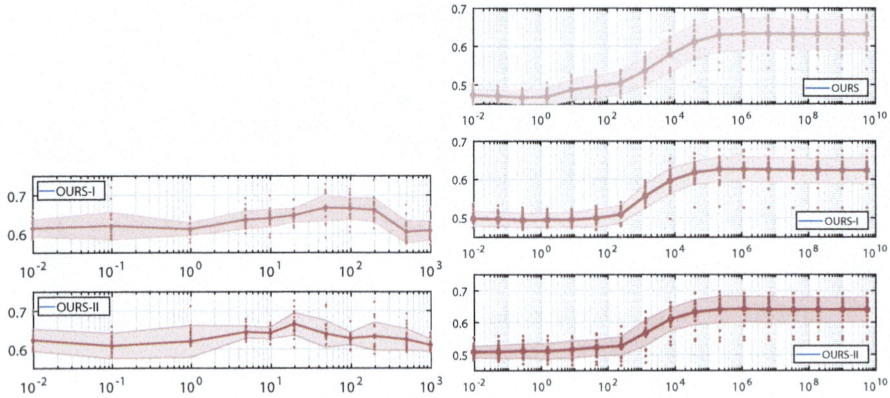

(a) (b)

Fig. 9.9 The identification performance variation on TE when varying **a** parameter R, **b** parameter γ

Parameter Study

Finally, we investigate how different parameters affect diagnostic performance, focusing on the aggressiveness parameter R for the PA algorithm and the trade-off parameter γ for updating the transformation matrix. Initially, we fix R and explore the range of γ from $[10^{-2}, 10^{10}]$. Subsequently, with γ fixed, we vary R within $[10^{-2}, 10^{3}]$. The variations in diagnostic performance under these different settings are presented in Fig. 9.9. Generally, satisfactory results are achieved when R is in the range of $[10^{0}, 10^{2}]$ and $\lg \gamma$ exceeds 4.

9.4.2 *Multi-phase Flow Facility*

Data Description

In this chapter, a normal class and two fault classes collected from the MPF facility are chosen for evaluation. Specifically, 22,966 samples collected across three operational modes are used for offline training. An additional 7,742 samples, acquired under two different modes, are initially used for testing and subsequently fed into the model for online updating. The testing results are reported for evaluation purposes. A detailed description of the dataset can be found in the Table 9.3.

Comparison

Consistent with the previous study, five non-adaptive and incremental models were selected for comparison, with settings aligned with those mentioned earlier. As shown in Fig. 9.10, the proposed algorithm outperforms the others in fault identification performance. The best results were achieved by OURS-I, with an accuracy

Table 9.3 The description of the selected MPF dataset

Description	Historical data			New data	
WFR (kg/s)	Varying	2	3.5	2	3.5
AFR (m^3/s)	Varying	0.0417	0.0208	0.0278	0.0417
Normal	7171	1347	1316	1321	1285
Fault 1	3616	3120	–	3000	–
Fault 2	4216	–	2180	–	2136
Total			**22966**		**7742**

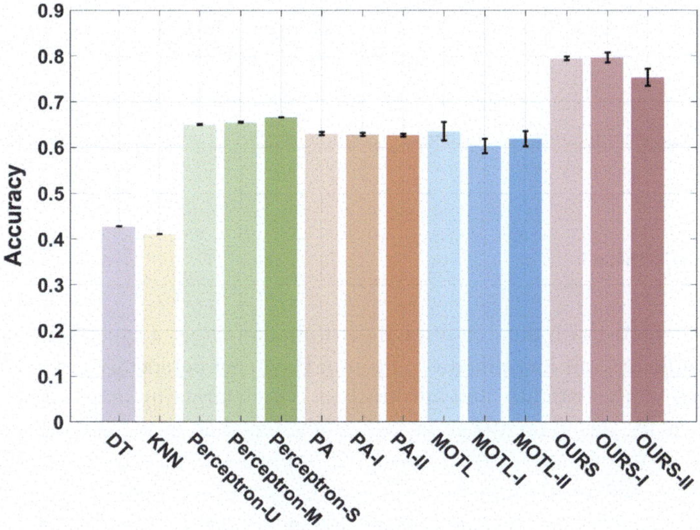

Fig. 9.10 The comparison results of different methods on the MPF

of 79.65%±1.05%. Compared to static models, all online models show greater effectiveness, as they can incrementally learn from newly received samples and adapt to the evolving environment. Additionally, the proposed algorithm significantly improves the performance of the standard PA algorithm, likely due to its distribution adaptation strategy. These findings are consistent with those observed in the prior experiments.

Parameter Study

We proceed by examining the sensitivity of the model to key parameters. Figure 9.11a shows the diagnostic performance for different values of R, while Fig. 9.11b illustrates the results for varying γ. Similar to the previous case, the results are satisfactory when R is within the range of $[10^0, 10^2]$ and $\lg \gamma$ falls between [3, 6]. These findings further confirm the robustness and efficiency of the proposed model across different parameter settings.

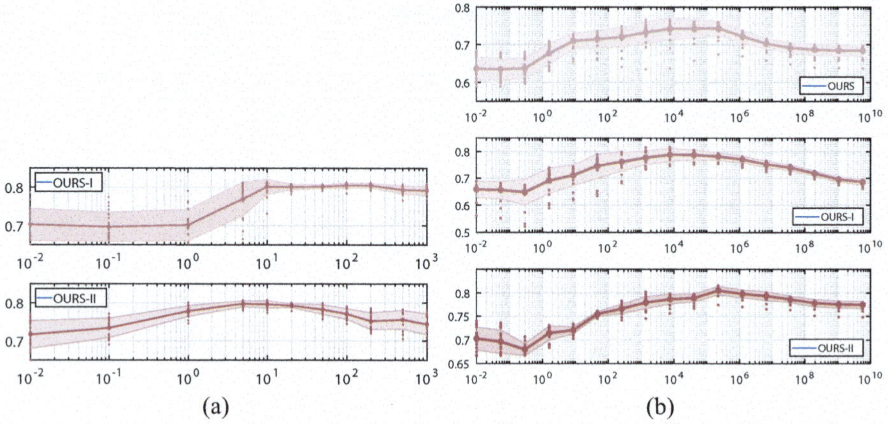

Fig. 9.11 The identification performance variation on MPF when varying **a** parameter R, **b** parameter γ

9.5 Conclusion

This chapter focuses on the conditional drift phenomenon, a specific aspect of distribution shift, in fault identification. From a Bayesian perspective, we define conditional drift mathematically and introduce an online targeting mapping strategy for adapting to drift. Based on this strategy, we developed an incremental identification model with drift adaptation capabilities. The proposed method is applied to both the TE benchmark and a real multi-phase flow process, and compared with several batch and incremental models. The results show that incremental models outperform batch models due to their ability to continuously learn from new data. Furthermore, our model demonstrates significant performance improvements over other incremental models, highlighting the effectiveness of drift adaptation in adjusting to the evolving behaviors of industrial systems. This makes our approach well-suited for practical scenarios where industrial data change over time and incremental learning is required.

This chapter uses the PA algorithm as the base classifier for our incremental identification model. However, in all scenarios, various pattern classification techniques, such as incremental support vector machines and online gradient descent, can be integrated into our framework. Additionally, we believe that incorporating non-linear methods, such as kernel functions and neural networks, as base classifiers could further improve the identification performance.

References

1. Seera M, Lim CP (2013) Online motor fault detection and diagnosis using a hybrid fmm-cart model. IEEE Trans Neural Netw Learn Syst 25(4):806–812
2. Mao W, He L, Yan Y, Wang J (2017) Online sequential prediction of bearings imbalanced fault diagnosis by extreme learning machine. Mech Syst Signal Process 83:450–473
3. Zhixin H, Jiang P (2018) An imbalance modified deep neural network with dynamical incremental learning for chemical fault diagnosis. IEEE Trans Indus Electron 66(1):540–550
4. Zhang K, Schölkopf B, Muandet K, Wang Z (2013) Domain adaptation under target and conditional shift. In: International conference on machine learning, pp 819–827. PMLR
5. Zhang X-Y, Liu C-L, Suen CY (2020) Towards robust pattern recognition: a review. Proc IEEE 108(6):894–922
6. Zhao P, Steven CH, Wang J, Li B (2014) Online transfer learning. Artif Intell 216:76–102
7. Qingyao W, Hanrui W, Zhou X, Tan M, Yonghui X, Yan Y, Hao T (2017) Online transfer learning with multiple homogeneous or heterogeneous sources. IEEE Trans Knowl Data Eng 29(7):1494–1507
8. Long M, Wang J, Ding G, Sun J, Yu PS (2013) Transfer feature learning with joint distribution adaptation. In: Proceedings of the IEEE international conference on computer vision, pp 2200–2207
9. Crammer K, Dekel O, Keshet J, Shalev-Shwartz S, Singer Y, Warmuth MK (2006) Online passive-aggressive algorithms. J Mach Learn Res 7(3):551–585
10. Freund Y, Schapire RE (1997) A decision-theoretic generalization of on-line learning and an application to boosting. J Comput Syst Sci 55(1):119–139
11. Borgwardt KM, Gretton A, Rasch MJ, Kriegel H-P, Schölkopf B, Smola AJ (2006) Integrating structured biological data by kernel maximum mean discrepancy. Bioinformatics 22(14):e49–e57
12. Crammer K, Singer Y (2003) Ultraconservative online algorithms for multiclass problems. J Mach Learn Res 3:951–991
13. Kang Z, Yang B, Li Z, Wang P (2019) Otlamc: an online transfer learning algorithm for multi-class classification. Knowl Based Syst 176:133–146
14. Ng A et al (2011) Sparse autoencoder. CS294A Lecture notes 72:1–19

Chapter 10
A Unified Fault Diagnosis Framework for Industrial Processes with Incremental Learning Ability

Many approaches often handle fault detection and identification tasks separately and require retraining when new fault samples or classes are introduced. To address this, we propose an incremental support vector data description scheme using a Gaussian kernel function, integrated within a unified framework for industrial fault diagnosis. This approach incrementally updates the decision boundary using only the original support vectors and newly acquired samples. To enhance detection accuracy, an adaptive threshold and restructured radius are introduced. For identification, hyperspheres representing known classes are constructed using a decision tree, with new samples that do not match any existing class being identified as belonging to an unknown class. This method enables simultaneous fault detection and identification without the need for time-consuming retraining. Experimental results validate the effectiveness and superiority of this incremental learning approach in fault diagnosis.

10.1 Motivation

While detection and identification problems have been explored in diagnostic tasks, most existing methods fail to simultaneously consider the adaptability of the decision boundary and the occurrence of new faults [1]. As a result, they do not enable the simultaneous and incremental execution of both tasks to incorporate newly collected fault data. As illustrated in Fig. 10.1, the industrial process initially operates stably, but fault samples (represented by red grids) accumulate, and new fault classes emerge over time. To address this challenge, this chapter proposes an SVDD-based fault diagnosis method with incremental capabilities tailored for industrial processes. This chapter aims to meet the critical need for more robust and unified fault diagnosis methods in the automation domain, particularly within industrial processes. The proposed method employs an adaptive threshold that is updated based on a specific subset of support vectors and a portion of newly collected samples. A restructured radius is introduced for the detection task. In the identification task, hyperspheres

H. Yin et al., *Data-Driven Fault Diagnosis for Complex Industrial Processes*,
Engineering Applications of Computational Methods 22,
https://doi.org/10.1007/978-981-96-3153-7_10

Fig. 10.1 Visual description of a simple industrial process

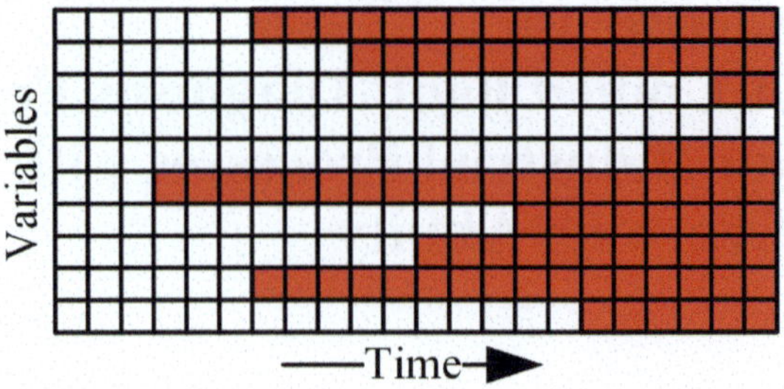

are constructed for each known fault class, and any sample that does not belong to an existing class is identified as belonging to a new class.

The contributions of this chapter are highlighted as follows:

- Industrial fault detection and identification can be achieved simultaneously on new data samples using an incremental approach.
- We propose a unified framework that allows fault diagnosis to be performed directly on newly collected data, without the need for dimensionality reduction or assumptions about data distribution.
- A decision tree-based ISVDD with an adaptive threshold is introduced to perform fault diagnosis tasks in kernel space.

10.2 Preliminaries Information

SVDD has been widely used across various fields, including anomaly detection, fault detection, identification, and pattern recognition [2]. Its goal is to construct an optimal region by identifying the minimum-volume hypersphere or domain that encompasses all or most of the objects. Given a training set $[x_1, x_2, \ldots, x_k] \in R^d$, where k represents the number of samples and d the dimensionality of each sample, the hypersphere with center c and radius r can be determined by solving the following problem:

$$\begin{aligned} &\min_{c,r} \ r^2 \\ &s.t. \ \ \|x_i - c\|^2 \le r^2, \forall i = 1, 2, \ldots, k. \end{aligned} \tag{10.1}$$

To improve fault tolerance and robustness, the slack variable ξ is typically introduced, allowing the model to more effectively adapt to data variations and noise [3]. Consequently, the optimization problem is reformulated as follows:

$$\begin{aligned} &\min_{c,r,\xi} \ r^2 + C\sum_{i=1}^{k} \xi_i \\ &s.t. \ \ \|x_i - c\|^2 \le r^2 + \xi_i, \xi_i \ge 0, \forall i = 1, 2, \ldots, k \end{aligned} \tag{10.2}$$

where the penalty parameter C balances misclassified samples rate and the volume of the hypersphere. For $C > 1$, the shape of the hypersphere remains unchanged.

By applying the Lagrange multiplier method and the KKT complementarity condition [4], the problem is transformed into its corresponding dual form, as shown in Eq. (10.3).

$$\begin{aligned} &\min_{\alpha_i} \sum_{i=1}^{k}\sum_{j=1}^{k} \alpha_i\alpha_j(x_i \cdot x_j) - \sum_{i=1}^{k} \alpha_i(x_i \cdot x_i) \\ &s.t. \quad 0 \le \alpha_i \le C, \sum_{i=1}^{k} \alpha_i = 1, \forall i = 1, 2, \ldots, k \end{aligned} \tag{10.3}$$

where α is the Lagrange operator. Given the typically high dimensionality of the kernel space, it is more efficient to compute the kernel function directly rather than performing the inner product operation. To achieve greater flexibility and accuracy, the kernel function $K(x_i, x_j) = \phi(x_i) \cdot \phi(x_j)$ replaces the inner product $(x_i \cdot x_j)$. The mapping function $\phi(\cdot)$ is implicitly defined through the kernel function, allowing the originally non-linear problem in low-dimensional space to be transformed into a linear problem in high-dimensional space [5]. Due to its high flexibility and ease of parameter optimization, the Gaussian kernel is the most widely used among kernel functions, as it effectively fits most functions with appropriate kernel parameters. Consequently, to enhance the performance of SVDD, the Gaussian kernel function $K(x_i, x_j) = \exp\left(-\frac{\|x_i - x_j\|^2}{\delta^2}\right)$ is generally employed.

Data samples are classified into three categories based on the values of the Lagrange multipliers:

$$\begin{cases} \|x_i - c\|^2 < r^2 \Rightarrow \alpha_i = 0 \\ \|x_i - c\|^2 = r^2 \Rightarrow 0 < \alpha_i < C \\ \|x_i - c\|^2 > r^2 \Rightarrow \alpha_i = C \end{cases} \tag{10.4}$$

A sample x with $\alpha > 0$ is referred to as a support vector, forming the support vector set V. Consequently, c and r are defined as follows:

$$c = \sum_{i=1}^{k} \alpha_i x_i \tag{10.5}$$

$$r = \sqrt{1 + \sum_{i=1}^{k}\sum_{j=1}^{k} \alpha_i\alpha_j K(x_i, x_j) - 2\sum_{i=1}^{k} \alpha_i K(x_n, x_i)} \tag{10.6}$$

where $x_n \in V$ and satisfies $0 < \alpha_n < C$. The distance d between any testing sample x_m and c is expressed as:

$$d = \sqrt{1 + \sum_{i=1}^{k}\sum_{j=1}^{k} \alpha_i \alpha_j K(x_i, x_j) - 2\sum_{i=1}^{k} \alpha_i K(x_m, x_i)} \tag{10.7}$$

If $d < r$, then x_m lies within the hypersphere and is considered part of the target object.

10.3 Methodology

This section outlines the research methodology. The Incremental Support Vector Data Description (ISVDD) technique is described in Sect. 10.3.1. Fault detection and identification are covered in Sect. 10.3.2. A unified framework for fault diagnosis is introduced in Sect. 10.3.3.

10.3.1 Incremental Support Vector Data Description

As industrial equipment—such as conveyor belts, sensors, and other machinery—operates, it continuously generates new data, including novel fault samples and classes. Traditional SVDD methods require retraining with the entire historical dataset whenever new data is introduced, resulting in a time-consuming process. Conversely, models that do not incorporate these continuously collected samples tend to underperform. Thus, ISVDD is essential for achieving more precise diagnostics.

In the conventional ISVDD algorithm, the model is updated using the support vectors from the previous model along with all newly collected samples. However, not all new samples violate the KKT condition; only those that do are useful for model updating. Additionally, when new samples do violate the KKT condition, the status of previous support vectors may change—some may become nonsupport vectors [6], while others might become new support vectors. To address this, an ISVDD algorithm with an adaptive radius has been proposed to enhance diagnostic accuracy.

In this chapter, during the incremental update step, the newly collected data consist of abnormal (fault) samples that deviate from expected industrial process behavior. If the distance d between a new sample and the center r of the previous model is greater than or equal to r, the sample falls outside the hypersphere defined by the support vectors. This indicates that the decision boundary is sufficiently robust and does not require updating. Conversely, if d is smaller than r, the hypersphere contains both normal and abnormal samples, suggesting that the decision boundary is no longer suitable. In this context, the relationship between distance and radius serves as an update trigger mechanism. When triggered, the boundary is adjusted based on the previous support vectors that satisfy $0 < \alpha < C$ and the newly collected samples where $d < r$.

Assume that the previous model, which includes the support vectors x and the radius r, is available. The support vectors from the previous model that satisfy $0 < \alpha < C$ are represented as $x = [x_1, x_2, \ldots, x_v]$. The new samples that meet the condition $d < r$ are denoted as $x' = [x_1', x_2', \ldots, x_l']$. Consequently, the updated training set can be expressed as $X = [x, x']$.

An adaptive threshold R, as defined in Eq. (10.8), is introduced to enhance the performance of the updated ISVDD model.

$$R = \frac{l}{N} r' + \frac{s}{N} r + \lambda \tag{10.8}$$

where l and s are the numbers of samples that meet and do not meet $d < r$, respectively.Let N denote the number of newly collected samples. The variables r and r' represent the radii of the previous and updated SVDD models, respectively. The parameter λ controls the tradeoff between the fault detection rate (FDR) and the fault alarm rate (FAR).

$$FDR = 1 - \frac{n_f^n}{n_f} \times 100\% \tag{10.9}$$

$$FAR = \frac{n_n^f}{n_n} \times 100\% \tag{10.10}$$

where n_f and n_n are the numbers of abnormal and normal samples, respectively. The terms n_f^n and n_n^f represent the number of fault samples incorrectly identified as normal, and the number of normal samples incorrectly identified as fault, respectively.

In Eq. (10.8), $\frac{l}{N}$ denotes the proportion of new samples that satisfy the previous KKT condition. Consequently, $\frac{l}{N} r'$ reflects the current model's ability to describe the new sample set, while $\frac{s}{N} r$ indicates the capability of the previous model. The threshold R varies depending on the training and testing sets, and it falls between r and r'.

The full procedure of the ISVDD algorithm is outlined in Algorithm 1.

Algorithm 1: ISVDD

Input: The original sample set and the newly collected sample set.

Output: An SVDD classifier M with incremental capabilities.

Step 1: Train the SVDD model M_0 using the original sample set, and calculate the hypersphere radius r.

Step 2: Obtain the new training set X based on support vectors set x and the new samples set x'. If $x' = \emptyset$, then set M_0 as M and end the incremental process.

Step 3: Train the SVDD model based on X to obtain the new hypersphere radius.

Step 4: Calculate R with Eq. (10.8) based on r and r'.

Step 5: End the incremental process.

10.3.2 Fault Detection and Identification

To enhance fault detection and minimize the impact of noise, the outputs are restructured. The jth restructured distance value, $d_{res}(j)$, is defined as follows:

$$d_{res}(j) = \sqrt{\sum_{t=1}^{q} \frac{t}{\sum_{t'=1}^{q} t'} d_{new}^2(j - q + t)} = \sqrt{\sum_{t=1}^{q} \frac{2t}{q(q+1)} d_{new}^2(j - q + t)} \quad (10.11)$$

where $d_{new}^2 = 1 + \sum_{i=1}^{v+l} \sum_{j=1}^{v+l} \alpha_i \alpha_j K(x_i, x_j) - 2 \sum_{i=1}^{v+l} \alpha_i K(x_m, x_i)$ is the distance between the testing sample x_m and the new hypersphere center c', q is the window length, and $x_i, x_j \in X$. Thus, a fault is indicated when $R < d_{res}$.

For fault diagnosis, fault detection alone is insufficient; identifying the specific fault type is also crucial. During the identification task, hyperspheres for all classes must be established through training. This process is guided by a decision tree-based SVM, which involves training multiple identification models. In a identification task with m classes, the training set is divided into positive and negative samples. The negative samples are further divided into two subcategories, and this subdivision continues until all training samples are classified. Consequently, $m - 1$ classifiers are required for a task involving m classes. For instance, to determine the hypersphere for class 1, the samples from class 1 are treated as positive samples, while samples from the remaining $m - 1$ classes are considered negative. This results in the construction of the classifier SVDD1. Next, the samples from class 2 are used as positive samples, and the samples from the other $m - 2$ classes (excluding classes 1 and 2) are treated as negative. This creates the classifier SVDD2. This process is repeated until the SVDD$m - 1$ classifier is constructed.

For any testing sample x_m, the distance d between x_m and each class center is compared with the corresponding radius to determine the class to which x_m belongs. If $d < r_1$, then x_m is classified as belonging to the first class. If not, the sample is then checked to see if it belongs to the second class, and so on, until all testing samples are successfully classified. If d exceeds all the radius values, the testing sample does not belong to any known class, indicating the presence of a new class.

10.3.3 The Overall Framework

The proposed incremental model with an adaptive threshold provides a unified framework for both fault detection and identification. For fault detection, the model leverages previous support vectors and newly collected samples to introduce an adaptive threshold, allowing for incremental updates without requiring a time-consuming retraining process. For fault identification, hyperspheres are constructed for each known fault class, turning the fault diagnosis task into multiple identification problems. A decision tree-based scheme is employed to guide the model in distinguishing between fault classes. New samples that do not fit into any known class are treated as representing a new class, thus enabling effective fault diagnosis even when new faults arise.

The fault diagnosis process framework is illustrated in Fig. 10.2. This comprehensive process includes both fault detection and fault identification. During the training phase, two models are established: an initial detection model and a identification model. In the initial detection model, z-score normalization is applied to mitigate the impact of outliers across different dimensions of the raw samples. This normalization is expressed as $x_{nor} = \frac{x-\mu(x)}{\delta(x)}$, where $\mu(x)$ and $\delta(x)$ represent the mean and standard deviation of the raw data, respectively. After normalization, the mean and standard deviation of x_{nor} are 0 and 1, respectively. In the identification model, hyperspheres for each fault class are created based on all labeled training samples. In the fault diagnosis phase, when new samples are introduced, an update mechanism is activated. The ISVDD-based detection model is incrementally and adaptively updated using an adaptive threshold R. Concurrently, the identification model is updated incrementally using a decision tree scheme. This process determines the decision boundary for each class, diagnoses known faults, and identifies new faults. Specifically, newly recorded samples are input directly into the well-trained ISVDD model to assess whether a fault has occurred. If a fault is detected, these samples are then processed by the trained identification model to determine the specific fault.

10.4 Case Study

10.4.1 Tennessee Eastman Process

This section provides a detailed analysis and discussion of the fault diagnosis performance, which is experimentally validated using the TEP benchmark. The selected TEP dataset includes various fault classes, each with different causes and associated variables. These faults are time-variant, and their data distributions fluctuate, making the TEP a useful benchmark for evaluating the effectiveness of the proposed method. Among the TEP faults, faults 3, 9, and 15, labeled as F3, F9, and F15, are incipient faults, characterized by minor deviations from normal operating conditions. These faults are subtle and their initial features are often barely noticeable. As a result, this chapter focuses on diagnosing the first 15 fault classes, excluding faults 3, 9, and 15.

Parameter Optimization

To achieve optimal performance, it is crucial to fine-tune the hyperparameters of the proposed model, as they significantly influence the results. In this chapter, the hyperparameters include the kernel parameter δ, the penalty parameter C, and the parameter λ for the adaptive threshold R. The kernel parameter δ controls the shape

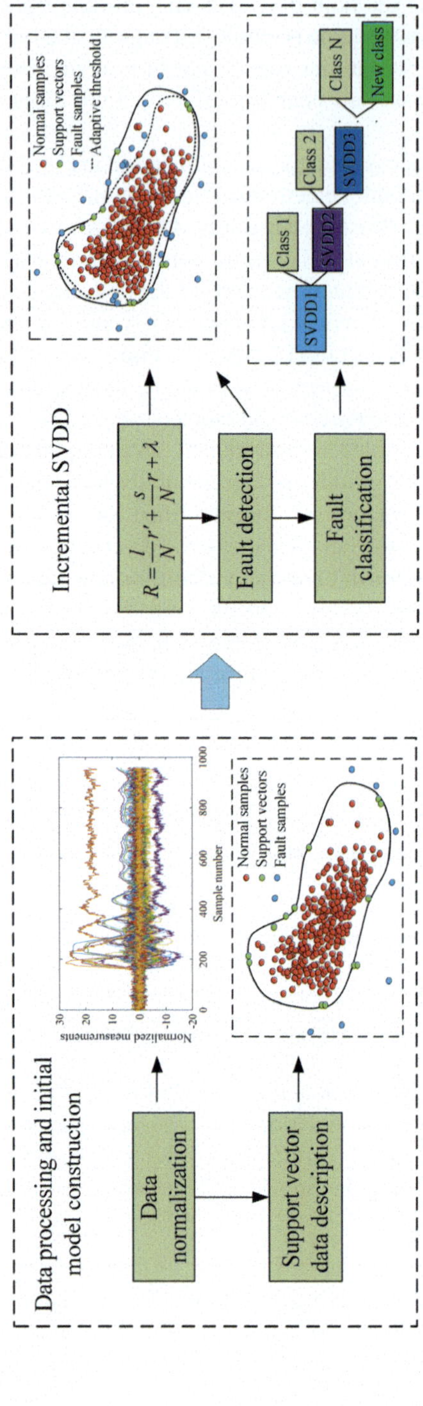

Fig. 10.2 The overall process of the proposed method for fault diagnosis

of the hypersphere. A larger δ results in a looser hypersphere boundary, so δ should be as small as possible. The penalty parameter C balances the rate of misclassified samples with the hypersphere volume. A larger C increases the hypersphere volume, which can lead to overfitting if set too high. Therefore, C should be minimized while still maintaining accuracy. The parameter λ manages the tradeoff between the FDR and the FAR, aiming to maximize FDR while minimizing FAR. The initial ranges for selecting these hyperparameters are set as follows: δ ranges from [7], C ranges from $[0.01, 0.1]$, and λ ranges from $[-0.004, 0.004]$.

The hyperparameters were optimized during the initial training process, which involved using all 500 normal samples and 160 samples from each fault class. Figure 10.3a, b illustrate the trends of the average fault detection rate (FDR) and average fault alarm rate (FAR) as functions of the kernel parameter δ and the penalty parameter C, respectively. Both the average FDR and average FAR decrease as δ increases, with minimal sensitivity to changes in C. Specifically, when δ exceeds 13, the average FDR drops below 90%, while δ values less than 11 result in an average FAR greater than 10%. To achieve a balance between FDR and FAR while keeping δ and C as small as possible, δ is set to 12 and C to 0.02. Figure 10.3c shows the variation of the average FDR and FAR with λ, which is set to -0.002 to optimize the tradeoff between these two rates.

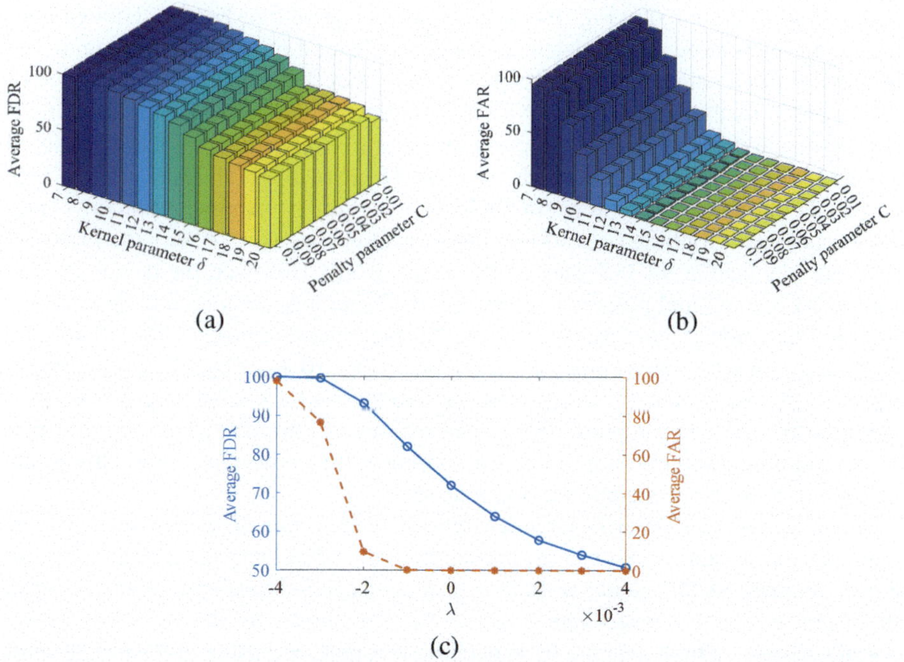

Fig. 10.3 Parameter optimization. **a** Average FDR varies with kernel parameter and penalty parameter. **b** Average FAR varies with kernel parameter and penalty parameter. **c** Average FDR and FAR vary with λ

Table 10.1 Indexes of incremental update and direct training for fault F1

Indexes	Figure 10.4a	Figure 10.4b	Figure 10.4c	Figure 10.4d
FDR (%)	99.38	99.38	99.38	99.38
FAR (%)	8.13	2.50	8.13	2.50
Time (s)	3.27	4.64	10.64	45.22

Fault Diagnosis Results

In this subsection, we validate the effectiveness of our approach through incremental experiments that incorporate both new fault samples and new fault classes.

Incremental Capability with New Samples

Fault Detection Results: The initial model is trained using 500 normal samples and the first 160 samples from each fault class in the training set. To simulate incremental learning with new samples, 40 samples from each fault class are sequentially introduced into the model during each incremental phase until all 480 abnormal samples per class are utilized for training. Each experimental trial tests the model using all 960 samples of each class in the testing set. For instance, consider fault F1. Figure 10.4a, b illustrate the results from the 1st and 8th incremental processes, respectively. Fault F1 is detected in the 166th observation sample. As shown in Fig. 10.4b, the FAR is lower compared to Fig. 10.4a, due to the adaptive threshold (red dotted line) increasing with more accumulated data. Figure 10.4c, d present results from direct training with all samples. A comparison reveals that the detection performance of the proposed method is consistent with that of direct training. The performance metrics are detailed in Table 10.1. Detection times for five repeated trials were measured using MATLAB 9.10.0 on a PC with Windows 10, an Intel i7-7500 CPU (2.70 GHz), and 8 GB of memory. As shown in Table 10.1, in the 1st incremental process (Fig. 10.4a) and direct training with 200 samples per fault class (Fig. 10.4c), the FDR and FAR are identical at 99.38% and 8.13%, respectively. However, the detection time for the 1st incremental process is shorter, at 3.27 seconds, compared to direct training. Similar findings are observed for the 8th incremental process (Fig. 10.4b) and direct training with 480 samples per fault class (Fig. 10.4d). Overall, while achieving the same performance at corresponding time points for F1, the proposed method significantly reduces computational cost compared to retraining. This suggests that incremental model updates are more time-efficient than retraining, especially as the number of new samples continues to increase.

Comparison with Other Detection Methods: To evaluate the effectiveness of our approach, we compare it with conventional ISVDD, KISVDD [6], NISVDD [7], and a fully retained SVDD model without incremental capabilities. Figure 10.5 displays the average detection performance across these methods. As shown in Fig. 10.5a, SVDD achieves a high average FDR because it is retrained using all historical samples. However, the performance of ISVDD declines as the number of training samples increases, likely due to overfitting since it does not account for the mapping between training and testing samples. Compared to other methods, the proposed approach

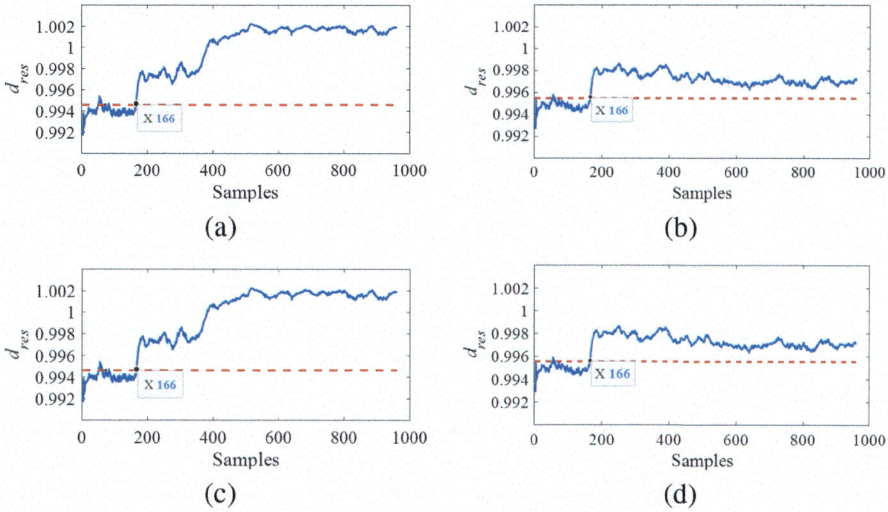

Fig. 10.4 Results of incremental update and direct training for F1. **a** The 1st incremental process. **b** The 8th incremental process. **c** Direct training with 200 samples per fault class. **d** Direct training with 480 samples per fault class

generally maintains the highest average FDR, except in the 1st incremental process where NISVDD performs better. However, this high FDR in NISVDD comes at the cost of an average fault alarm rate (FAR) of 77.42%, which is too high to be graphically represented. Figure 10.5b shows that the average FAR for our method ranges from a minimum of 0.78% to a maximum of 7.03% across the 1st and 6th incremental processes. Although the average FAR for the proposed method is relatively higher in most cases, it tends to decrease when the number of training samples per class is below 440. In the 7th incremental process, the average FAR increases, likely because the distances of newly collected samples from the center of the previous hypersphere are smaller than the corresponding radius. This results in a smaller radius for the new hypersphere to maintain a higher average FDR. A similar issue leads to high average FARs in NISVDD and SVDD when the training samples per class are 440.

Additionally, Fig. 10.6 illustrates the average detection times for various methods. It is evident that as the number of training samples increases, SVDD and KISVDD require progressively more time. In contrast, methods with incremental capabilities are significantly faster than their non-incremental counterparts, primarily because incremental methods update the model using only new samples, thereby reducing the overall time cost. Our proposed method consistently achieves an average detection time of approximately 4 seconds across all tasks. Overall, the proposed approach not only performs fault detection efficiently but also excels in time-effectiveness.

To further validate the effectiveness and superiority of our approach, we compare it with several competitive methods: KPCA [8], reduced interval KPCA (RIKPCA)

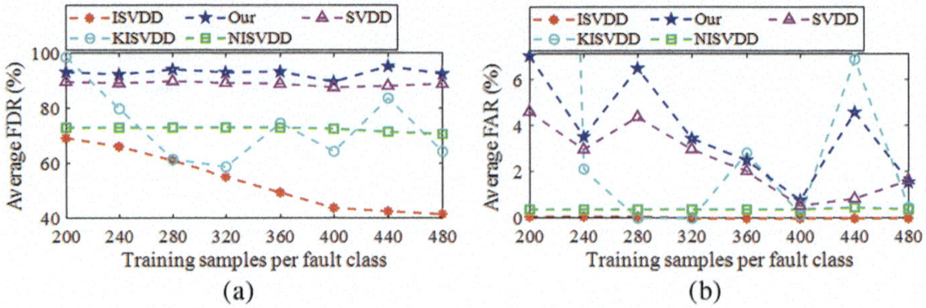

Fig. 10.5 Detection performance of different methods. **a** Average FDR. **b** Average FAR

Fig. 10.6 Average time consumption of different methods

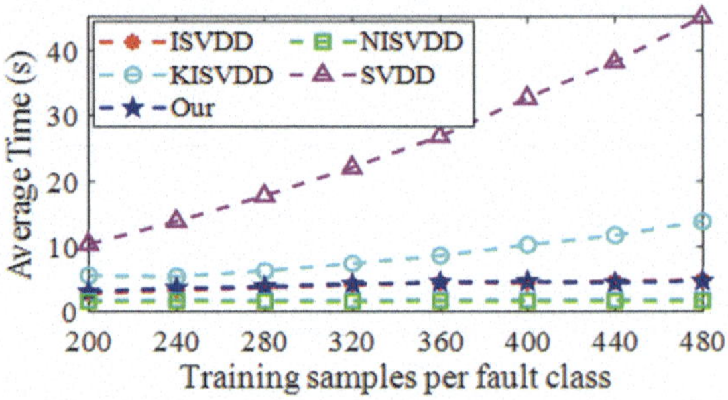

[8], SVDD [9], stacked autoencoder (SAE) [10], part interval SAE-SVDD (PISAE-SVDD) [10], DCAA [11], variable selection SVDD (VSSVDD) [12], and MI-Louvain-DSVDD [13]. Note that these methods are trained using all available samples from each fault class. The FDR and FAR values for these methods are summarized in Tables 10.2 and 10.3, respectively. These metrics together assess the fault detection capabilities. According to Table 10.2, our method achieves the highest average FDR, while RIKPCA performs the worst. Table 10.3 indicates that our method

Table 10.2 FDR (%) of different methods for the first 8 faults in the TEP

	RIKPCA	KPCA	P_σ	SVDD	SAE-SVDD	PISAE-SVDD	VSSVDD	DCAA	MI-Louvain-DSVDD	Our
F1	98.84	100.0	91.25	99.38	99.40	99.40	99.00	98.75	99.75	99.38
F2	98.58	99.22	98.66	98.63	98.40	98.50	98.00	96.88	98.63	98.50
F4	2.06	47.80	99.80	99.38	100.0	100.0	100.0	95.88	100.0	99.25
F5	21.00	64.17	79.82	47.38	74.00	75.00	57.00	95.88	33.00	77.00
F6	100.0	100.0	87.95	99.88	99.70	99.70	100.0	100.0	100.0	99.90
F7	38.65	76.03	78.01	100.0	100.0	100.0	100.0	84.13	100.0	100.0
F8	85.96	100.0	87.71	97.88	97.90	98.10	98.00	86.50	98.00	97.75
F10	29.38	92.91	91.36	69.75	39.50	44.00	73.00	44.88	59.75	68.88
Average	59.31	85.02	89.32	89.04	88.61	89.34	90.63	87.86	86.14	92.58

Table 10.3 FAR (%) of different methods for the first 8 faults in the TEP

	RIKPCA	KPCA	P_σ	SVDD	SAE-SVDD	PISAE-SVDD	VSSVDD	DCAA	MI-Louvain-DSVDD	Our
F1	0	26.87	3.27	2.50	5.60	4.40	0	1.88	3.13	2.50
F2	0.89	23.21	12.11	0.63	1.30	0.60	0	0	1.88	0.63
F4	1.78	24.55	0.10	1.25	1.30	0.06	3.00	5.00	2.50	0
F5	1.78	24.55	21.38	1.25	4.40	4.40	3.00	6.25	1.88	2.50
F6	0.44	20.53	13.23	0	4.40	1.80	0	2.50	3.75	0
F7	0.89	24.55	16.25	0	3.10	1.90	2.00	1.88	3.13	0
F8	0.89	41.07	28.5	6.88	3.80	1.30	0	1.25	6.25	6.25
F10	1.78	21.87	14.49	0.63	5.00	2.50	1.00	3.13	2.50	0.63
Average	1.06	25.90	13.67	1.64	3.61	2.12	1.13	2.74	3.13	1.56

ranks third with an average FAR of 1.56%. The lowest average FAR is achieved by RIKPCA at 1.06%. Despite some classes showing lower FDR compared to other methods, our approach excels in FAR for specific classes, such as F4, F6, F7, and F10. The difference between our method's average FAR and the optimal average FAR is only 0.50%. Overall, these results confirm the effectiveness and superiority of our proposed method with an adaptive threshold.

Fault identification Results: Upon detecting a fault, the identification mechanism is activated to determine the fault type. For example, consider fault F4. The identification results, as shown in Fig. 10.7, indicate that most fault samples for F4 are accurately predicted. Figure 10.8 presents the multiclass fault identification confusion matrix, where fault samples that do not match any known fault class are labeled as the unknown fault class (0). From the confusion matrix, it is observed that fault F10 has the lowest identification accuracy among all faults, at 68.12%. However, the majority of faults are classified correctly, with F1 and F8 performing particularly well. Overall, the average identification accuracy across all faults is 92.16%. These results demonstrate that the proposed method delivers reliable and satisfactory fault identification performance.

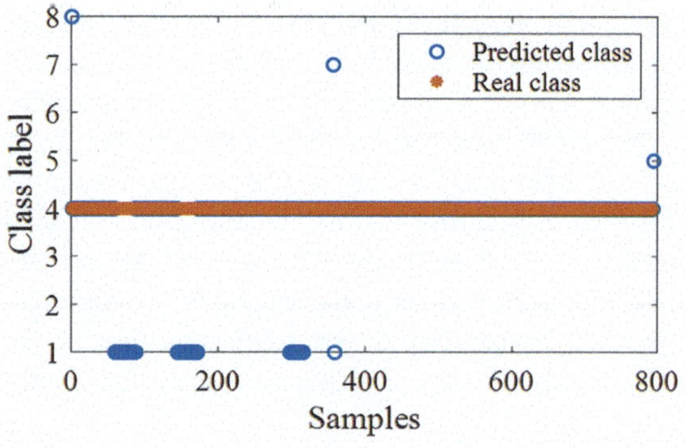

Fig. 10.7 The predicted classes for detected fault samples of F4

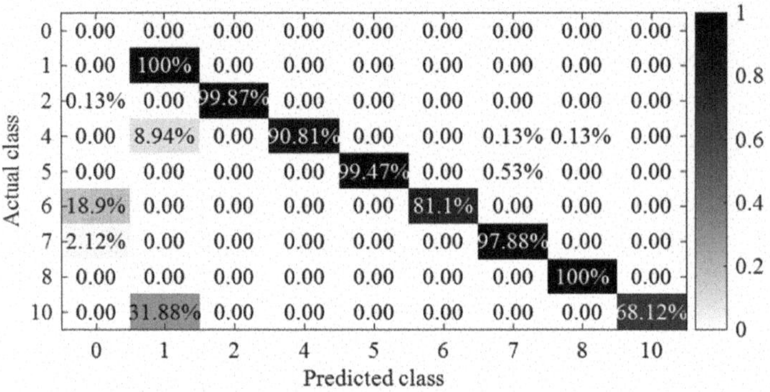

Fig. 10.8 Confusion matrix for detected fault samples

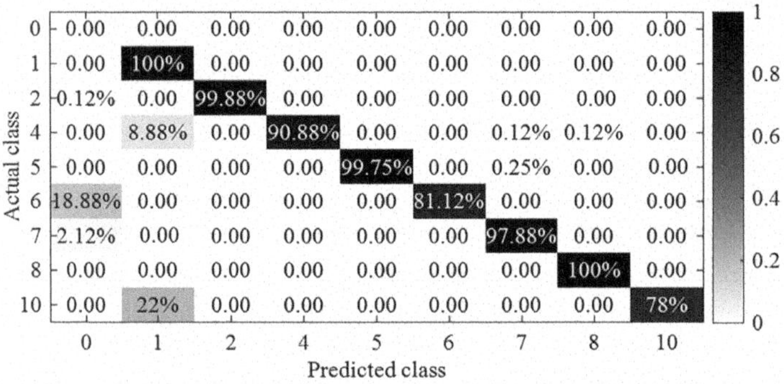

Fig. 10.9 Confusion matrix for all fault samples in each class

Comparison with Other identification Methods: To further evaluate the identification accuracy of our method, we tested all 800 abnormal samples from each fault class in the TEP testing set. The results, shown in Fig. 10.9, demonstrate that our method achieves high fault identification accuracy, with most fault classes exceeding 90%. Notably, F1 and F8 have a perfect accuracy of 100%. Additionally, we compared our method with several competitive approaches, including SVM [14], ANN, SAE, deep belief network (DBN), BLS [15], CNN [16], and BCNN [17]. Table 10.4 presents the identification rates and time consumption for each method, with optimal values highlighted in bold. According to the table, the average identification rate for all methods is above 75%, with our method achieving the highest rate. Specifically, our method's average identification rate is 7.06% higher than that of ANN, the best competitor in this regard. In terms of time consumption, our method takes 24.07 seconds, ranking third among all methods. Although SVM and BLS have lower time consumption, their identification rates are around 80%. An effective method should balance minimal time consumption with high identification accuracy. Thus, our method proves to be highly effective for diagnosing various faults in the TEP.

Table 10.4 Identification rates (%) and time consumption (s) of all methods for all methods the first 8 faults in the TEP

	SVM	ANN	SAE	DBN	BLS	CNN	BCNN	Our
Class 1	97.00	98.33	98.75	98.50	98.50	97.61	98.75	100.0
Class 2	96.00	97.50	98.00	97.38	98.50	99.24	98.00	99.88
Class 4	97.12	91.46	99.12	99.75	99.00	90.83	99.75	90.88
Class 5	96.63	97.5	99.25	99.25	98.12	97.26	100.0	99.75
Class 6	44.75	99.79	51.00	49.75	72.75	87.39	76.00	81.12
Class 7	100.0	100.0	100.0	100.0	99.88	99.73	100.0	97.88
Class 8	60.88	92.08	49.50	55.13	53.75	54.16	57.87	100.0
Class 10	39.87	14.38	37.75	41.50	29.88	27.13	34.00	78.00
Average	79.03	86.38	79.17	80.16	81.30	81.67	83.05	93.44
Time	14.085	70.825	83.300	63.582	20.523	191.147	29.376	24.070

Incremental Capability with New Fault Classes

Diagnosis Performance Analysis: This subsection further validates the diagnostic performance of our method. Initially, the model is trained using the first 8 faults listed in Table 10.1. Subsequently, a new fault class is introduced into the well-trained model at each step until all four fault classes are included. Figure 10.10 presents the diagnosis performance for the initial 8 faults. Panel (a) shows the average fault detection rate (FDR) and fault alarm rate (FAR). It is evident that the average FDR increases gradually as additional fault classes are introduced. However, the average FAR rises significantly with the inclusion of F13, a new fault type characterized by slow drift, which reduces the decision boundary size. Similarly, the addition of another new fault type, F14, further reduces the boundary, resulting in an increased average FAR. Panel (b) of Fig. 10.10 displays the identification performance for the initial 8 faults in the presence of new fault classes. The confusion matrix indicates that the overall identification performance remains satisfactory, with particularly strong results for F1 and F8.

Identification of New Fault Classes: this chapter also evaluates the identification performance for new fault classes (F11-F14). The results are depicted in Fig. 10.11a, showing identification accuracies of 87.75%, 95.37%, 44.12%, and 100%, respectively. Figure 10.11b provides a comparison of average accuracy across methods. It is evident that the average accuracy for the new classes using our proposed method significantly surpasses that of other methods. Additionally, the detection and identification times for these new fault classes are 4.87, 5.12, 4.87, and 4.86 seconds, respectively. These results demonstrate that our method delivers satisfactory diagnostic performance for new fault classes.

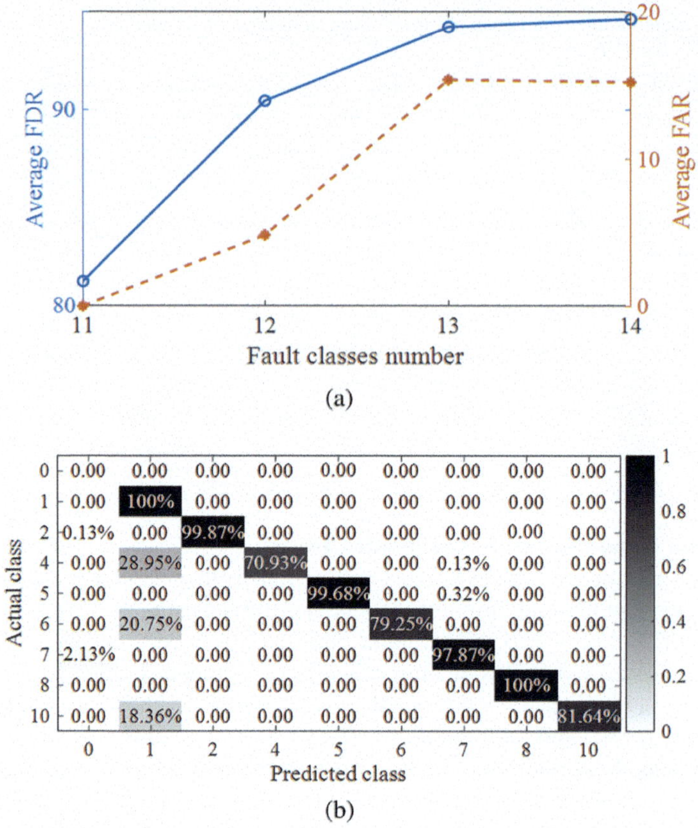

Fig. 10.10 The diagnosis performance of the first 8 faults. **a** The average FDR and FAR. **b** The identification performance

10.5 Conclusion

This chapter presents a unified incremental framework with an adaptive threshold for fault diagnosis in industrial processes. This framework allows for adaptive and incremental updates directly with new data, eliminating the need for complete retraining when new samples or classes are introduced. A key advantage of this approach is its ability to perform both fault detection and multilabel identification simultaneously, without relying on dimensionality reduction or distributional assumptions. The adaptive threshold balances the tradeoff between fault detection rate (FDR) and fault alarm rate (FAR). For fault identification, the method employs hypersphere-shaped boundaries for each fault class using a decision tree-based strategy, enabling the diagnosis of new classes. Experimental results demonstrate that the proposed method effectively updates with new fault samples and classes, offering both time efficiency and superior diagnostic performance.

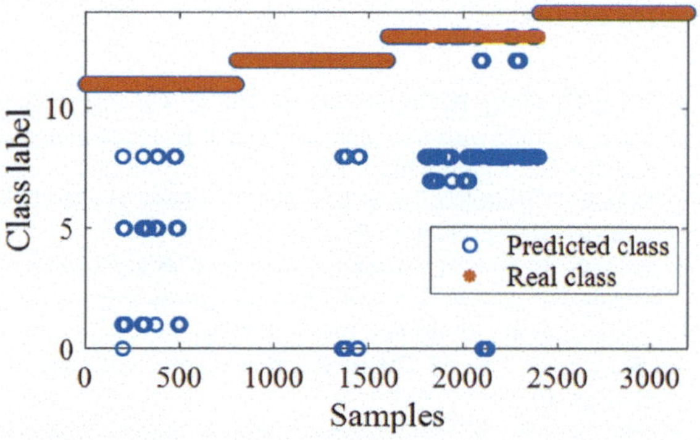

(a)

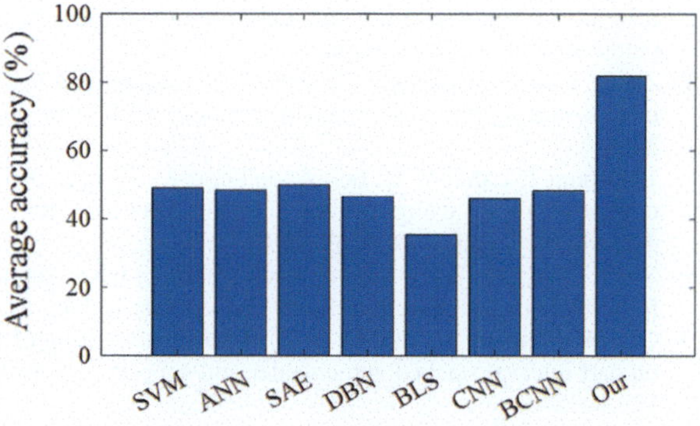

(b)

Fig. 10.11 The diagnosis results and average accuracy for the new fault classes F11-F14. **a** The diagnosis results of the proposed method. **b** Comparisons with other methods

Like many studies, the proposed method has some limitations. For instance, the experiments are conducted using a public benchmark dataset, and the approach assumes an equal allocation of new samples across all fault classes during the incremental process. Similarly, when introducing new classes, it assumes an equal number of samples for each new fault class and only adds one new fault class per incremental step. In real-world applications, obtaining samples for specific faults can be challenging, and new fault classes often have significantly fewer samples compared to existing classes, potentially leading to class imbalance issues. Additionally, practical scenarios may involve the simultaneous occurrence of multiple faults, which could impact performance. Future work will focus on validating the proposed method with more realistic processes and developing a robust fault diagnosis framework that addresses class imbalance and multiple simultaneous faults to improve diagnostic accuracy.

References

1. Jiang H, Wang H, Wenhao H, Kakde D, Chaudhuri A (2019) Fast incremental svdd learning algorithm with the gaussian kernel. In Proceedings of the AAAI conference on artificial intelligence 33, pp 3991–3998
2. Sohrab F, Raitoharju J, Iosifidis A, Gabbouj M (2021) Multimodal subspace support vector data description. Pattern Recogn 110:107648
3. Jingjing L, Gao Y, Zhang L, Deng H, Cao J, Bai J (2022) A novel dynamic radius support vector data description based fault diagnosis method for proton exchange membrane fuel cell systems. Inte J Hydrogen Energy 47(84):35825–35837
4. Wang C-D, Lai JH (2013) Position regularized support vector domain description. Pattern Recogn 46(3):875–884
5. Hou H, Ji H (2021) Improved multiclass support vector data description for planetary gearbox fault diagnosis. Control Eng Pract 114:104867
6. Yin G, Zhang Y-T, Li Z-N, Ren G-Q, Fan H-B (2014) Online fault diagnosis method based on incremental support vector data description and extreme learning machine with incremental output structure. Neurocomputing 128:224–231
7. Zhang L, Qiao F, Wang J, Zhai X (2019) Equipment health assessment based on improved incremental support vector data description. IEEE Trans Syste Man Cybern Syst 51(5):3205–3216
8. Hamrouni I, Lahdhiri H, ben Abdellafou K, Taouali O (2020) Fault detection of uncertain nonlinear process using reduced interval kernel principal component analysis (rikpca). Inte J Adv Manuf Technol 106:4567–4576
9. Yin L, Wang H, Fan W (2018) Active learning based support vector data description method for robust novelty detection. Knowl Based Syst 153:40–52
10. Qiqi W, Weipeng L, Yan X (2022) Process monitoring of nonlinear uncertain systems based on part interval stacked autoencoder and support vector data description. Appl Soft Comput 129:109570
11. Chen Z, Cao Y, Ding SX, Zhang K, Koenings T, Peng T, Yang C, Gui W (2019) A distributed canonical correlation analysis-based fault detection method for plant-wide process monitoring. IEEE Trans Indus Inform 15(5):2710–2720
12. Cai L, Yin H, Lin J, Zhou H, Zhao D (2022) A relevant variable selection and svdd-based fault detection method for process monitoring. IEEE Trans Autom Sci Eng 20(4):2855–2865
13. Wang J, Liu P, Shan L, Zhou M, Chen X (2023) Decentralized plant-wide monitoring based on mutual information-louvain decomposition and support vector data description diagnosis. ISA Trans 133:42–52
14. Chang C-C, Lin C-J (2011) Libsvm: a library for support vector machines. ACM Trans Intell Syst Technol 2(3):1–27
15. Philip Chen CL, Liu Z (2017) Broad learning system: an effective and efficient incremental learning system without the need for deep architecture. IEEE Trans Neural Netw Learn Syst 29(1):10–24
16. Hao W, Zhao J (2018) Deep convolutional neural network model based chemical process fault diagnosis. Comput Chem Eng 115:185–197
17. Wanke Yu, Zhao C (2019) Broad convolutional neural network based industrial process fault diagnosis with incremental learning capability. IEEE Trans Indus Electron 67(6):5081–5091

The manufacturer's authorised representative in the EU is Springer Nature Customer Service Centre GmbH, Europaplatz 3, 69115 Heidelberg, Germany. If you have any concerns regarding our products, please contact ProductSafety@springernature.com

Printed and bound by CPI Group (UK) Ltd, Croydon, CR0 4YY
01/10/2026
02225937-0001